T0329557

CONTROL OF CUTTING VIBRATION AND MACHINING INSTABILITY

CONTROL OF CUTTING VIBRATION AND MACHINING INSTABILITY

A TIME-FREQUENCY APPROACH FOR PRECISION, MICRO AND NANO MACHINING

C. Steve Suh and Meng-Kun Liu
Texas A&M University, USA

A John Wiley & Sons, Ltd., Publication

Registered office
John Wiley & Sons Ltd, The Atrium, Southern Gate, Chichester, West Sussex, PO19 8SQ, United Kingdom

For details of our global editorial offices, for customer services and for information about how to apply for permission to reuse the copyright material in this book please see our website at www.wiley.com.

Library of Congress Cataloging-in-Publication Data

Suh, C. Steve
   Control of cutting vibration and machining instability : a time-frequency approach for precision, micro and nano machining / C. Steve Suh and Meng-Kun Liu.
      pages cm
   Includes bibliographical references and index.
   ISBN 978-1-118-37182-4 (cloth)
   1. Cutting–Vibration.   2. Machine-tools–Vibration.   3. Machining.   4. Machinery, Dynamics of.
5. Time-series analysis.   6. Microtechnology   7. Nanotechnology.   I. Liu, Meng-Kun.   II. Title.
   TJ1186.S86 2013
   671.3'5–dc23

                                                                                          2013014158

A catalogue record for this book is available from the British Library

ISBN: 9781118371824

Typeset in 10/12pt Times by Aptara Inc., New Delhi, India

Printed and bound in Singapore by Markono Print Media Pte Ltd

Contents

Preface

Controllers are either designed in the frequency domain or time domain. When designed in the frequency domain, it is a common practice that a transfer function is derived from the corresponding governing equation of motion. Frequency response design methods, such as bode plots and root loci, are usually employed for the design of frequency domain based controllers. When designed in the time domain, the differential equation of the system is described as a state space model by the associated state variables. The controllability and observability of the design are then investigated using state feedback or other time domain control laws. Controllers designed in either domain have their advantages. Controllers designed in the frequency domain provide adequate performance with uncertainties. Estimating the output phase and amplitude in response to a sinusoidal input is generally sufficient to design a feedback controller, but the system has to be linear and stationary. Controllers designed in the time domain can accommodate multiple inputs-outputs and correlate internal and external states without considering the requirements in the frequency domain. Proven feasible for linear, stationary systems, both types of controllers can only be applied independently either in the frequency domain or the time domain.

For a linear time-invariant system, only the amplitude and phase angle of the input are changed by the system. The output frequency remains the same as the input frequency, and the system can be stabilized by applying a proper feedback gain. Both the time domain and frequency domain responses are bounded at the same time. However, this is not the case for the chaotic response generated by a nonlinear system. A chaotic response is naturally bounded in the time domain while becoming unstably broadband in the frequency domain, containing an infinite number of unstable periodic orbits of all periods in the phase portrait, called strange attractors. It does not remain in one periodic orbit but switches rapidly between these unstable periodic orbits. If the chaotic response is projected onto the Poincaré section, a lower dimensional subspace transversal to the trajectory of the response, numerous intersection points would be seen densely congregating and being confined within a finite area. This unequivocally implies that the chaotic response is bounded within a specific range in the time domain and dynamically deteriorates at the same time with a changing spectrum of broad bandwidth as the trajectory switching rapidly between infinite numbers of unstable periodic orbits. This phenomenon is prevalent in high-speed cutting operations where strong nonlinearities including regenerative effects, frictional discontinuity, chip formation, and tool stiffness are dominant.

For a nonlinear, nonstationary system, when it undergoes bifurcation to eventual chaos, its time response is no longer periodic. Broadband spectral response of additional frequency components emerge as a result. Controllers designed in the time domain contain time-domain error while being unable to restrain the increasing bandwidth. On the other hand, controllers designed in the frequency domain restrict the bandwidth from expanding while losing control over the time domain error. Neither frequency domain nor time domain based controllers are sufficient to address aperiodic response and route-to-chaos. This is further asserted by the uncertainty principle, which states that the resolution in the time- and frequency-domain cannot be reached arbitrarily. However, as Parseval's Theorem implies that the energies computed in the time- and frequency-domains are interchangeable, it is possible to incorporate and meet the time and frequency domain requirements together and realize the control of a nonlinear response with reconciled, concurrent time-frequency resolutions.

The above is a concise version of what was on our mind when we contemplated many years ago the following two questions: (1) *Why is it that a dynamic response could be bounded in the time domain while in the meantime becoming unstably broadband in the frequency domain simultaneously?* and (2) *Why is it that the control of a nonlinear system has to be performed in the simultaneous time-frequency domain to be viable and effective?* The wavelet-based time-frequency control methodology documented in this monograph is the embodiment of our response to these particular questions.

The control methodology is adaptive in that it monitors and makes timely and proper adjustments to improve its performance. Plant parameters in the novel control are identified in real-time and are used to adjust and update the control laws according to the changing system output. Its architecture is inspired by active noise control in which one FIR filter identifies the system and another auto-adjustable FIR filter rejects the uncontrollable input. Analysis wavelet filter banks are incorporated in the control configuration. The analysis filter banks decompose both input and error signals before the controlled signal is synthesized. As a dynamic response is resolved by the discrete wavelet transform into components at various scales corresponding to successive octave frequencies, the control law is inherently built in the joint time-frequency domain, thus facilitating simultaneous time-frequency control. Unlike active noise control whose objective is to reduce acoustic noise, the wavelet-based time-frequency control is designed to minimize the deterioration of the output signal in both the time and frequency domains when the system undergoes bifurcated or chaotic motion so as to restore the output response to periodic. These features together provide unprecedented advantages for the control of nonlinear, nonstationary system response.

This book presents a sound foundation which engineering professionals, practicing and in training alike, can rigorously explore to realize important progress in micro-manufacturing, precision machine-tool design, and chatter control. Viable solution strategies can be formulated drawing from the foundation to control cutting instability at high speed and to develop chatter-free machine-tool concepts. Research professionals in the general areas of nonlinear dynamics and nonlinear control will also find the volume informative in qualitative and quantitative terms on how discontinuity and chaos can be adequately mitigated.

The discourse of *Control of Cutting Vibration and Machining Instability* is organized into eleven chapters. The first chapter examines the coupled tool–workpiece interaction for a better understanding of the instability and chatter in turning operation. The second chapter is a brief review of the mathematical basics along with the common notations relevant to the derivation of the wavelet-based nonlinear time-frequency control law in Chapter 7. The third chapter reviews the essences of active noise control and the filtered-x LMS algorithm that are incorporated in the time-frequency control as features for system identification and error reduction. The notion of time-frequency analysis is discussed in the fourth chapter. This chapter presents several analysis tools important for the proper characterization of nonlinear system responses. It also lays the fundamentals needed to comprehend wavelet filter banks and the underlying concept of time-frequency resolution that are treated in the chapter that follows, Chapter 5.

The philosophical basis on which the nonlinear time-frequency control is based is elaborated in greater detail in Chapter 6. Chapter 7 presents the time-frequency control theory along with all the salient physical features that render chaos control feasible. The feasibility is demonstrated in applications in Chapters 8 through 11 using examples from high-speed manufacturing and friction-induced discontinuity. The last chapter, Chapter 11, explores an alternative solution to mitigating chaos using the time-frequency control. The implication for exploring synchronization of chaos to achieve suppressing self-sustained chaotic machining chatter is emphasized.

Two working MATLAB® *m*-files by which all the results and figures in Chapters 10 and 11 are generated are listed in the Appendix. The one for the friction-induced instability control in Chapter 10 has an extensive finite element coding section in it that utilizes several user-defined MATLAB® functions in Simulink® for the calculation of the beam vibration. We hope the examples will encourage the gaining of practical experience in implementing the wavelet-based nonlinear time-frequency control methodology. Readers who are reasonably familiar with the MATLAB® language and Simulink simulation tool should find the examples extensive, complete, and easy to follow.

Since it was first conceived years ago, many talented individuals have come along and helped evolve the core ideas of time-frequency control. Among them are Baozhong Yang, who explored instantaneous frequency as the tool of preference for characterizing nonlinear systems, and Achala Dassanayake, to whom we owe the comprehensive understanding of what machining instability and chatter really are.

C. Steve Suh and Meng-Kun Liu
Texas A&M University, College Station, USA
February 2013

1
Cutting Dynamics and Machining Instability

Material removal – as the most significant operation in manufacturing industry – is facing the ever-increasing challenge of increasing proficiency at the micro and nano scale levels of high-speed manufacturing. Fabrication of submicron size three-dimensional features and freeform surfaces demands unprecedented performance in accuracy, precision, and productivity. Meeting the requirements for significantly improved quality and efficiency, however, are contingent upon the optimal design of the machine-tools on which machining is performed. Modern day precision machine-tool configurations are in general an integration of several essential components including process measurement and control, power and drive, tooling and fixture, and the structural frame that provides stiffness and stability. As dynamic instability is inherently prominent and particularly damaging in high-speed precision cutting, *design for dynamics* is favored for the design of precision machine tool systems [1]. This approach employs computer-based analysis and design tools to optimize the dynamic performance of machine-tool design at the system level. It is largely driven by a critical piece of information – the vibration of the machine-tool. Due to the large set of parameters that affect cutting vibrations, such as regenerative effects, tool nonlinearity, cutting intermittency, discontinuous frictional excitation, and environmental noise, among many others, the effectiveness of the approach commands that the dynamics of machining be completely established throughout the entire process.

This book explores the fundamentals of cutting dynamics to the formulation and development of an innovative control methodology. The coupling, interaction, and evolution of different cutting states are studied so as to identify the underlying critical parameters that can be controlled to negate machining instability and enable better machine-tool design for precision micro and nano manufacturing.

The main features that contribute to the robust control of cutting instability are: (1) comprehension of the underlying dynamics of cutting and interruptions in cutting motions, (2) operation of the machine-tool system over a broad range of operating conditions with

Control of Cutting Vibration and Machining Instability: A Time-Frequency Approach for Precision, Micro and Nano Machining, First Edition. C. Steve Suh and Meng-Kun Liu.
© 2013 John Wiley & Sons, Ltd. Published 2013 by John Wiley & Sons, Ltd.

minimal vibration, such as high-speed operation to achieve a high-quality finish of the machined surface, (3) an increased rate of production to maximize profit and minimize operating and maintenance costs, (4) concentration on the apparent discontinuities that allows the nature of the complex machine-tool system motions to be fully established. The application of simultaneous time-frequency nonlinear control to mitigate complex intermittent cutting is both novel and unique. The impact on the area of material removal processes is in the mitigation of cutting instability and chaotic chattering motion induced by frictional and tool nonlinearity, and (5) development of concepts for cutting instability control and machine-tool design applicable to high-speed cutting processes.

1.1   Instability in Turning Operation

We start the book with a comprehensive investigation on machining instability by employing a three-dimensional turning model [2, 3, 4, 5] that addresses the concerns that (1) cutting dynamic models developed to date all fall short of grasping the underlying dynamics of turning operation and (2) stability charts developed using the models are inadequate to identify the true stability regions. The specific objective of the study is to establish the proper interpretation of cutting instability so as to establish the knowledge base for cutting instability control.

The complex machining model describes the coupled tool–workpiece dynamics subject to nonlinear regenerative cutting forces, instantaneous depth-of-cut, and workpiece whirling due to material imbalance. In the model the workpiece is considered a system of three rotor sections – namely, unmachined, being machined, and machined – connected by a flexible shaft, thus enabling the motion of the workpiece relative to the tool and tool motion relative to the machining surface to be three-dimensionally established as functions of spindle speed, depth-of-cut, rate of material removal, and whirling. Figure 1.1 shows the configuration of the tool engaging the section that is being cut where the deviation of the geometric center from the center of mass constitutes the eccentricity that characterizes workpiece whirling. Using the model a rich set of nonlinear behaviors of both the tool and workpiece – including period-doubling bifurcation and chaos signifying the extent of machining instability at various depth-of-cuts – was observed. Results presented therein agree favorably with physical

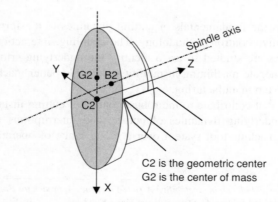

C2 is the geometric center
G2 is the center of mass

Figure 1.1   Configuration of the section that is being machined

experiments reported in the literature. It is found that at, and up to, certain ranges of depth-of-cuts, whirling is a non-negligible part of the fundamental characteristics of the machining dynamics. Contrary to common belief, whirling is found to have an insignificant impact on tool motions. The efforts documented in [2, 3, 4, 5] also show that the linearized turning model underestimates tool vibrations in the time domain and overestimates system behaviors in the frequency domain; whereas the nonlinear model agrees with the physical results reported in the literature in describing machining stability and chatter. The coupled workpiece–tool vibrations described by the nonlinear model are more pragmatic than the linearized counterpart in revealing the true machining state of motion. The model also reveals in the qualitative sense the broadband behavior of the tool natural frequency associated with unstable situations. Vibration amplitudes obtained using the linearized model, however, are diverging at certain depth-of-cuts (DOCs) without the commonly observed randomness in oscillation. Moreover, the linearized model deems instability at low DOCs and predicts a bifurcated state of unstable motion that is described as chaotic using physical data. Many important insights are gained using the model, including the fact that if the underlying dynamics of machining is to be established, and stability limits to be precisely identified, linearization of the nonlinear model is not advisable.

1.1.1   Impact of Coupled Whirling and Tool Geometry on Machining

In addition to speed, feed rate, and DOC that affect material removal rate (MRR) and determine cutting force and hence power consumption, tool geometry is also one of the prominent parameters that impacts machining productivity. Surface roughness, chip formation changes, and chip flow angle are also affected by tool geometry. Even though chip flow angle is related to tool angles, chip flow angle is a function only of DOC. Figure 1.2 gives a view of the rake angle, α, while undergoing cutting action. Tool rake angle determines the flow of the newly formed chip. Usually the angle is between $+5°$ and $-5°$. To compare with the experimental result reported in [6], a constant spindle speed $\Omega = 1250$ rpm, a constant chip width $t_0 = 0.0965$ mm, and an eccentricity $\varepsilon_1 = 0.2$ mm are considered along with several different DOCs including DOC = 1.62 mm and 2.49 mm. The workpiece considered is a 4140 steel bar of 0.25 m length (l_0) and $r_3 = 20.095$ mm radius of the machined section. The starting location of the carbide tool is set at 0.15 m from the chuck. There are three types of plots in the figures found in the sections that follow. The top rows plot time histories, whereas the middle rows give their corresponding time-frequency responses obtained using instantaneous frequency (which will be covered in great detail in Chapter 4). The last rows show the Lyapunov spectra where the largest Lyapunov exponents are shown. Instantaneous frequency is employed to realize subtle features characteristic of machining instability.

Positive rake makes the tool sharp, but it also weakens the tool compared with negative rake. Negative rake is better for rough cutting. The selection of tool geometry depends on the particular workpiece and tool materials being considered. To establish that tool angle does have significant effects on cutting stability, two sets of tool geometries are used to determine the cutting force in the following. Their values are given in Table 1.1. Both sets are taken from the tool inserts that were used in the experiments reported in [6]. Since DOC considered in the numerical study is less than 1 mm and can be considered as non-rough cutting, rake angles are taken as positive for all cases. Note that negative rake is better for roughing. Three

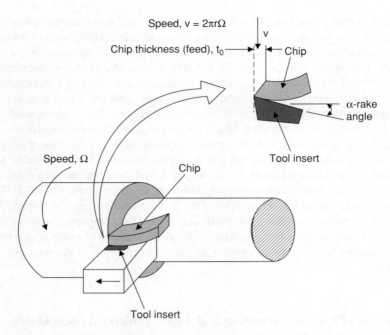

Speed, v = 2πrΩ

Chip thickness (feed), t_0

Chip

α-rake angle

Tool insert

Speed, Ω

Chip

Tool insert

Figure 1.2   Cutting action with tool rake angle

DOCs (0.9 mm, 0.75 mm, and 0.5 mm) are used with a 1250 rpm spindle speed and a feed of 0.0965 mm per revolution in the study.

Except for Figures 1.6 and 1.7, all figures in Figures 1.3–1.8 give time responses, instantaneous frequency responses between 3 to 5 sec, and the corresponding Lyapunov spectra. X-direction system responses are examined to demonstrate workpiece behaviors, and Z-direction responses are analyzed to investigate tool motions. See also Figure 1.1 for the coordinates defined for the being-machined section. Plots in the right column correspond to Set #1 tool geometry conditions and the left column corresponds to Set #2 tool angles. In Figure 1.3, the X-direction vibration amplitude of Set #2 tool geometry is seen to be twice that of Set #1. However, their frequency domain behaviors are similar with a workpiece natural frequency at 3270 Hz and a whirling frequency at 20.8 Hz. Set #2 shows two more frequencies, one near the tool natural frequency at 425 Hz and another at 250 Hz, which disappears after 3.9 s. Set #1 has only one more tool-excited frequency. The frequency can be seen to decrease from 580 Hz to 460 Hz within 2 seconds, implying that tool geometry is a non-negligible parameter affecting workpiece stability.

Table 1.1   Tool angles

Set Number	Side cutting edge angle	Rake angle	Inclination angle
1	45°	3.55°	3.55°
2	15°	5°	0°

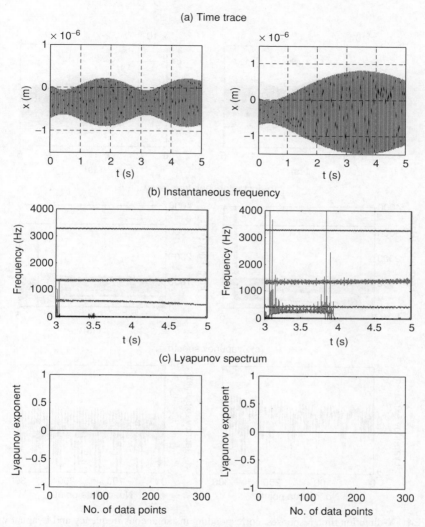

Figure 1.3 X-direction time responses, corresponding instantaneous frequency and Lyapunov spectra for Set #1(left) and Set #2 (right) for DOC = 0.90 mm at Ω = 1250 rpm

When DOC is decreased to 0.75 mm in Figure 1.4, there are still differences in vibration amplitudes. With the reduction of its diameter, the workpiece natural frequency decreases to 3250 Hz. While whirling frequency remains the same, a 900 Hz component of a wide 500 Hz bandwidth dominates in both systems. It can be seen in Set #2 that a bifurcation of the tool-excited natural frequency at 425 Hz diminishes after 4.8 seconds. On the other hand, Set #1 does not have a bifurcation. It has a frequency component increase from 250 to 400 Hz. The frequency components then disappear afterward. Both Lyapunov spectra fluctuate near zero, thus leaving a question over whether the systems are exactly stable. Further deceasing DOC

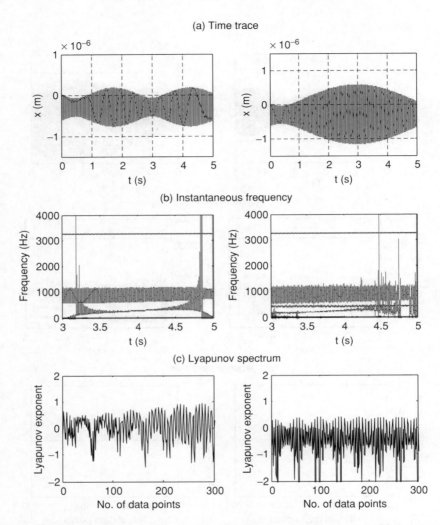

Figure 1.4 X-direction time responses, corresponding instantaneous frequency and Lyapunov spectra for Set #1(left) and Set #2 (right) for DOC = 0.75 mm at $\Omega = 1250$ rpm

to 0.5 mm in Figure 1.5 means both systems show an unstable situation marked by positive Lyapunov exponents.

The relatively large force fluctuation seen in Figure 1.6 explains the large vibration amplitudes seen for the tool geometry Set #2. Forces of large fluctuation push the workpiece to deflect more. It is interesting to note that, even though tool geometry variations are supposed to affect the cutting force, X-direction force amplitudes are almost identical for both tool geometry sets.

Effects of tool geometry can be seen in the Y- and Z-direction force components in Figure 1.7. While Y-direction forces for Set #2 are less than those of Set #1, Set #2 Z-direction forces are

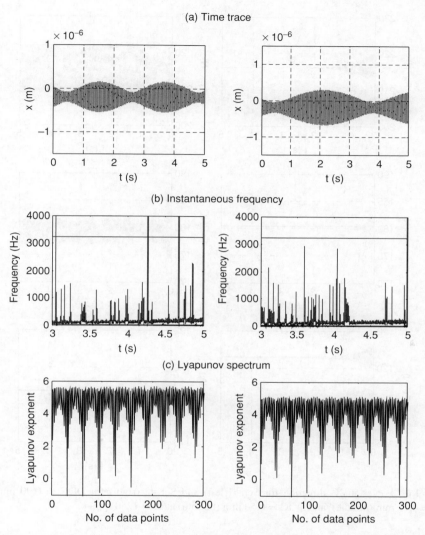

Figure 1.5 X-direction time responses, corresponding instantaneous frequency and Lyapunov spectra for Set #1(left) and Set #2 (right) for DOC = 0.50 mm at $\Omega = 1250$ rpm

much higher than those of Set #1. In all plots, it is seen that force responses associated with Set #1 tool geometry fluctuate less compared with those of Set #2. Tool dynamical motions for DOC = 0.9 mm, 0.75 mm, and 0.5mm are also considered. Even though force fluctuations and vibration amplitudes are both prominent, Set #2 is relatively more stable. Of the three DOCs considered, two behave dissimilarly. In all three cases, the vibration history of Set #1 has amplitudes that are of nanometers in scale. On the other hand, Set #2 vibrates with amplitudes that are a few microns for DOC = 0.9 mm and 0.75 mm, and several nanometers for DOC = 0.5 mm. Unlike Set #1, all Lyapunov spectra for Set #2 are evidence of a stable state of dynamic

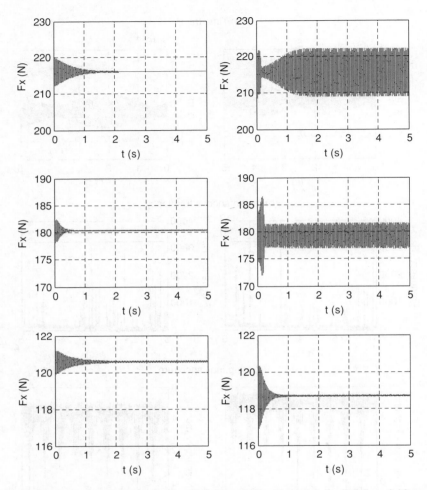

Figure 1.6 Forces in X- direction for Set #1(left) and Set #2 (right) for DOC = 0.90 mm (top), DOC = 0.75 mm (middle), and DOC = 0.50 mm (bottom) at Ω = 1250 rpm

response. Though it has positive Lyapunov exponents, Set #2 shows instability for the DOC = 0.9 mm and 0.5 mm cases. IF plots for DOC = 0.9 mm in Figure 1.8 confirm that the Set #1 response is broadband and thus unstable, and Set #2 is stable with a clean spectrum. Although the Lyapunov spectrum indicates a stable state of tool motion for Set #1 at DOC = 0.75 mm, the corresponding instantaneous frequency suggests otherwise. The instantaneous frequency plot for Set #2 at DOC = 0.5 mm also contradicts the Lyapunov spectrum (not shown). A detailed review of the individual instantaneous frequency mono-components reveals that the frequency at 3240 Hz has bifurcated three times. Thus, it is in a highly bifurcated state.

The effects of tool geometry on cutting dynamics and its impact on surface finishing investigated above generate a few observations. The manufacturing industry has long learned to employ tool inserts with complex geometry to achieve better product surface finish. However,

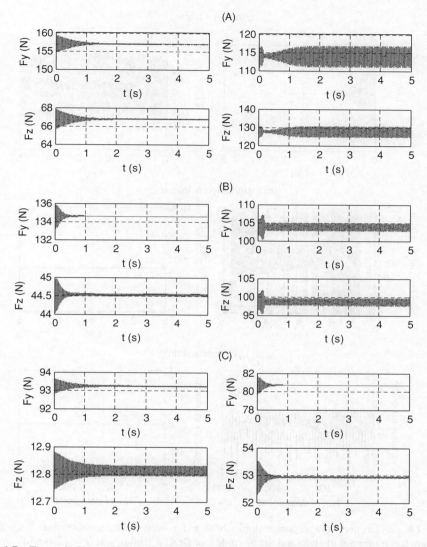

Figure 1.7 Forces in Y- and Z-direction for Set #1 (left) and Set #2 (right) for DOC = 0.90 mm (A), DOC = 0.75 mm (B), and DOC = 0.50 mm (C) at Ω = 1250 rpm

most models developed for understanding machining dynamics and cutting stability ignore the various effects attributable to tool geometry. One of the reasons for this is the fact that 1D machining modeling is infeasible for incorporating various tool angles that are inherently 3D. Numerical experiments presented suggest that neglecting tool geometry is improbable. It was observed that variations in tool geometry can significantly impact cutting stability. A machining process can be unstable for a particular DOC using one set of tool geometry and become stable through careful selection of proper tool inserts with a different set of tool

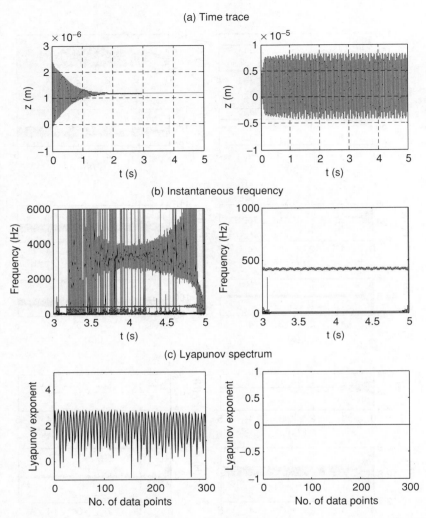

Figure 1.8 Z-direction time responses (tool vibrations), corresponding instantaneous frequency and Lyapunov spectra for Set #1 (left) and Set #2 (right) for DOC = 0.90 mm at $\Omega = 1250$ rpm

geometry at the same DOC. This raises the question over whether true dynamic stability can be identified without considering tool geometry. Thus, it is essential that tool geometry is also considered in modeling 3D turning operations.

1.2   Cutting Stability

The dynamic model is simulated using four different spindle speeds and several depth-of-cuts. Results presented in this section are only for DOC = 1.62 mm with a 0.0965 mm chip thickness.

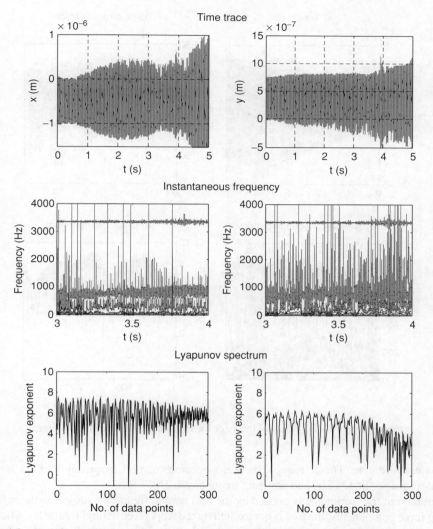

Figure 1.9   X-(left) and Y-(right) direction workpiece responses for DOC = 1.62 mm at Ω = 750 rpm

In what immediately follows, we consider the time responses of the workpiece and the tool. Two spindle speeds, 750 rpm and 1000 rpm, are considered that correspond to the same DOC at 1.62 mm. Vibrations in all directions in Figure 1.9 and 1.10 seem random and the positive Lyapunov spectra indicate chaos. These are accompanied by broadband instantaneous frequency spectra. When the speed is increased to 1000 rpm, it can be seen in Figures 1.11 and 1.12 that both the workpiece and tool responses are remarkably stable. The time traces are periodic and the Lyapunov spectra are zero, implying that future vibrations can be predicted. Moreover, the instantaneous frequency plots show only three major frequencies

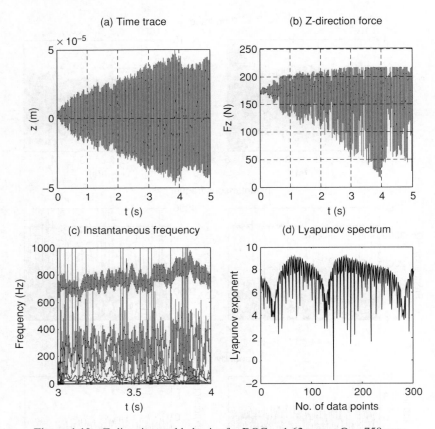

Figure 1.10   Z-direction tool behavior for DOC = 1.62 mm at $\Omega = 750$ rpm

that do not bifurcate. These frequencies are: workpiece natural frequency at 3343 Hz, tool natural frequency at 425 Hz, and whirling frequency at 16.67 Hz.

Even though the workpiece is excited by the tool natural frequency subject to the action of cutting force components, the tool is not excited by the workpiece natural frequency. These are very important observations made with regard to the two different spindle speeds. Although physically engaged with the workpiece, however, the dynamics that governs the tool response is distinctively different from that of the workpiece's. This has a significant implication for the interpretation of cutting stability and the judging and characterizing of chatter. Simply put, machining chatter could be due to not just the cutting tool but also the workpiece.

1.3   Margin of Stability and Instability

The stable–unstable margins for rough turning at large DOCs are found for four different speeds at 750 rpm, 1000 rpm, 1250 rpm, and 1500 rpm. Workpiece and tool vibration responses of the cutting model system that correspond to stable DOCs and unstable DOCs near the critical

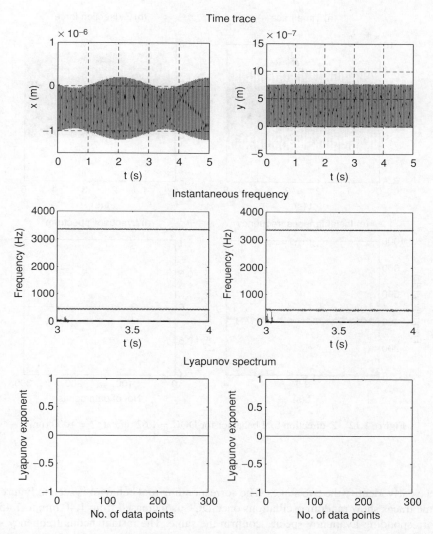

Figure 1.11   X-(left) and Y-(right) direction workpiece responses for DOC = 1.62 mm at $\Omega = 1000$ rpm

depth-of-cut are presented in the figures that follow. X- and Z-direction vibration responses are examined to identify the workpiece and tool stability margins respectively. Workpiece responses found in Figures 1.13, 1.14, 1.15, and 1.16, each illustrates a case of stability and a case of instability subject to two different DOCs at the same spindle speed. This is performed to identify the margin of instability for the particular DOCs considered.

Figure 1.13 shows the workpiece behaviors at close to the critical DOC at $\Omega = 750$ rpm. It is seen that the workpiece vibrates with stability subject to DOC = 1.40 mm, but becomes chaotic when DOC is increased to 1.45 mm at the same speed. Zero and positive Lyapunov

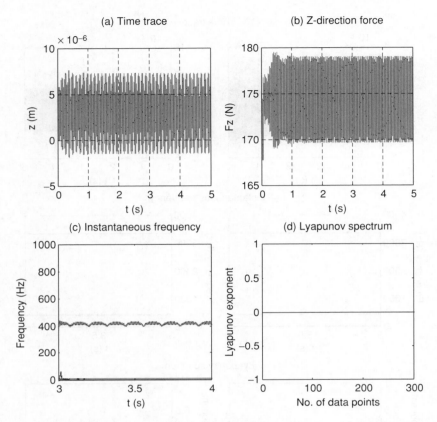

Figure 1.12   Z-direction tool behavior for DOC = 1.62 mm at $\Omega = 1000$ rpm

spectra clearly identify the two stages. The corresponding tool behavior is seen in Figure 1.17. The time traces show random oscillations once DOC is increased from 1.40 mm to 1.45 mm. The corresponding Lyapunov spectra confirm the same. The instantaneous frequency shows broadband characteristics in the lower frequency region for the stable case. The frequency excited by the tool natural frequency remains unvaried with time. Unlike the 1.40 mm case, when DOC is 1.45 mm the excited tool characteristic frequency varies in time. This frequency behavior can be seen in all the unstable plots of the tool in Figures 1.17, 1.18, 1.19, and 1.20 (right column) using large DOCs.

When speed is increased to 1000 rpm, the critical DOC also increases. Figure 1.14 shows the results of the workpiece vibrations corresponding to two different DOCs. It can be seen that the case corresponding to DOC = 1.75 mm is very stable in both the time and frequency domains. On the other hand, having a positive Lyapunov spectrum and broadband frequency characteristics, the case with DOC = 1.78 mm is an instability of chaos. The corresponding tool behaviors are found in Figure 1.18. Similar to the case with DOC = 750 rpm, violent oscillations, large vibration amplitudes, and a positive Lyapunov spectrum signify an unstable

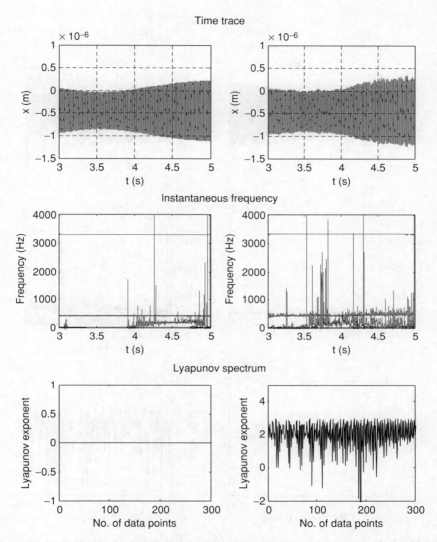

Figure 1.13 X-direction workpiece behavior for DOC = 1.40 mm (left) and DOC = 1.45 mm (right) at Ω = 750 rpm

state of tool motion at DOC = 1.78 mm. Note that the response associated with the DOC is only 0.03 mm less than 1.78 mm, which is a case of stability.

Increasing the spindle speed further to 1250 rpm results in a stable state of cutting, thus moving the instability margin up to the 1.83–84 mm range (Fig. 1.15). Again, the associated time history, instantaneous frequency, and Lyapunov spectrum of the tool all indicate a state of chaos for DOC = 1.84 mm. This is in contrast with the case of DOC = 1.83 mm whose zero Lyapunov spectrum indicates stability. However, the time-frequency characteristics of

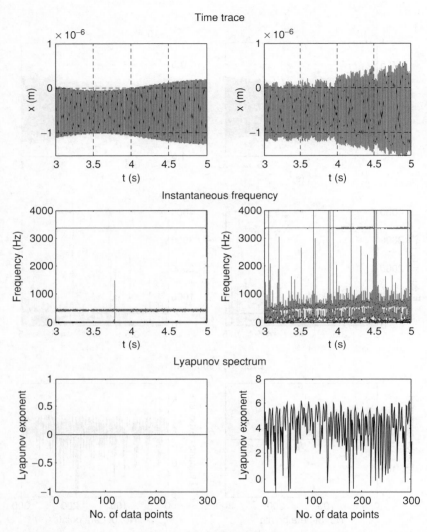

Figure 1.14 X-direction workpiece behavior for DOC = 1.75 mm (left) and DOC = 1.78 mm (right) at Ω = 1000 rpm

the DOC = 1.83 mm case reveal that it is a bifurcated state. An increment of only one hundredth of a millimeter is enough to tip the stable cutting motion into dynamic instability. The tool responses in Figure 1.19 for the two DOCs convey a slightly different characteristic than the vibration results associated with the workpiece. Tool vibration amplitude is only several nanometers for the case with DOC = 1.83 mm. The 0.01 mm in increment induces chaotic responses of 0.1 mm in vibration amplitude. The state of chaotic motion is verified by the Lyapunov spectrum in Figure 1.19. Unlike all other cases considered, the instantaneous

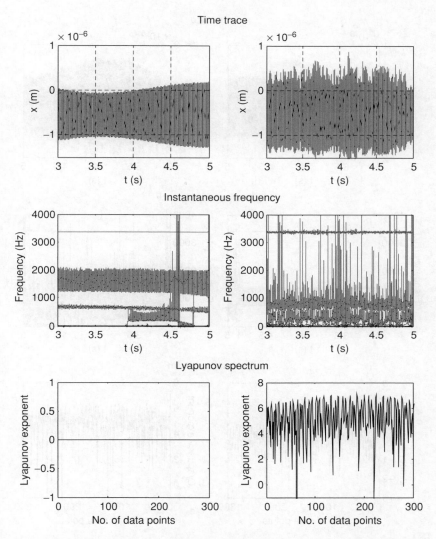

Figure 1.15 X-direction workpiece behavior for DOC = 1.83 mm (left) and DOC = 1.84 mm (right) at $\Omega = 1250$ rpm

frequency shows that the tool vibrates with a frequency closer to the workpiece natural frequency at the lower DOC. This frequency component immerges after 3.8 seconds. The Lyapunov spectrum of positive exponents for this case indicates that the motion is unstable.

Figure 1.16 shows the X-direction workpiece behaviors for DOC = 2.21 mm and 2.22 mm at $\Omega = 1500$ rpm. The time and frequency domain responses depict two different behaviors for the two DOCs. The Lyapunov spectrum for the lower DOC is not a straight line of zeros. It is an oscillation about zero, thus indicating that the motion is of a stable–unstable marginal

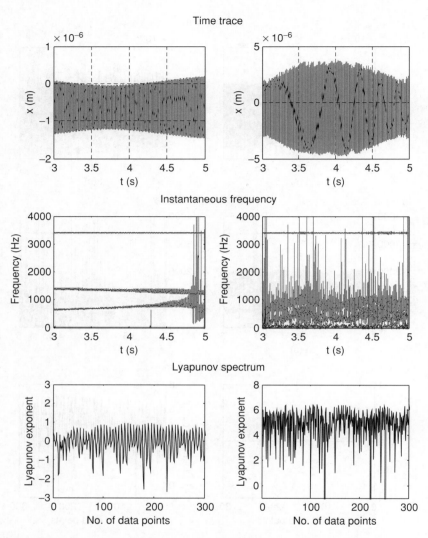

Figure 1.16   X-direction workpiece behavior for DOC = 2.21 mm (left) and DOC = 2.22 mm (right) at $\Omega = 1500$ rpm

type. The tool behavior at this speed is comparable to the case in Figure 1.20 where the speed was 1250 rpm. The vibration amplitude grows more than 10 000 times greater when DOC is increased by one hundredth of a millimeter from 2.21 mm to 2.22 mm. The instantaneous frequency for the lower DOC has a component of broadband characteristic at the workpiece natural frequency that appears after 3.5 seconds. The Lyapunov spectra for both DOCs are also positive, as is the case with $\Omega = 1250$ rpm. Using all the above cases of stable–unstable margin, the critical DOCs are determined for rough cutting at the various spindle speeds considered. Details will be given in the following section.

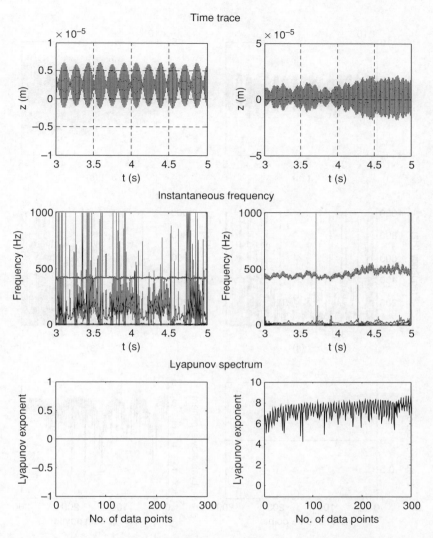

Figure 1.17 Z-direction tool behavior for DOC = 1.40 mm (left) and DOC = 1.45 mm (right) at $\Omega = 750$ rpm

Instead of spindle speed, cutting speed is the most common cutting parameter used in stability charts. For a workpiece having a machined section of D millimeters in diameter, the corresponding cutting speed is defined as follows

$$cutting\_speed = \frac{\pi D \Omega}{1000 \times 60} \ ms^{-1}$$

where Ω is spindle speed in revolutions per minute. This means that there can be different spindle speeds and workpiece diameters that have the same cutting speed. In milling or drilling,

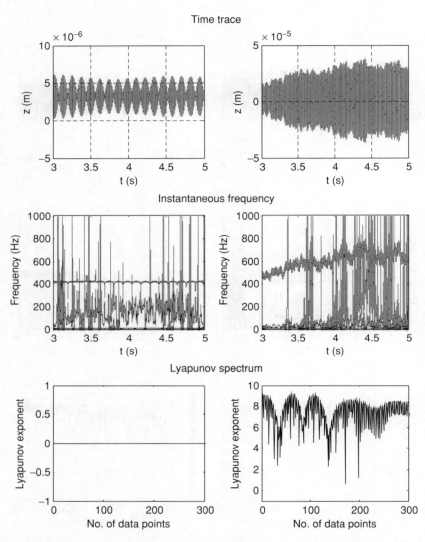

Figure 1.18 Z-direction tool behavior for DOC = 1.75 mm (left) and DOC = 1.78 mm (right) at $\Omega = 1000$ rpm

milling tools or drill bits have standard diameters. However, in turning processes, the machined diameter can be any value depending on the product requirements. The critical DOCs that were explored previously are compared with the experimental results in [7]. Experimentally found critical DOCs for cutting speeds in the range of 50 m/min to 300 m/min are published in [7].

Using the definition above, the stability limits can be presented as the critical DOC for each corresponding cutting speed as in Table 1.2.

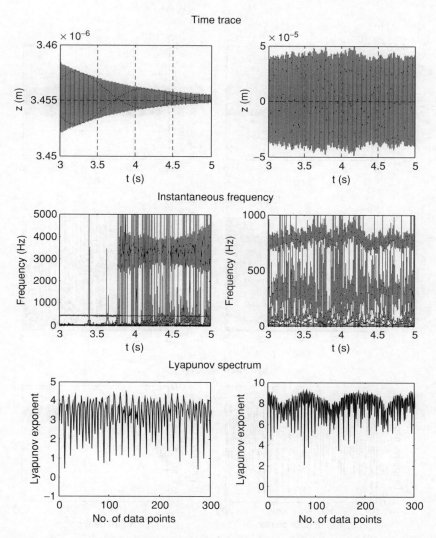

Figure 1.19 Z-direction tool behavior for DOC = 1.83 mm (left) and DOC = 1.84 mm (right) at $\Omega = 1250$ rpm

Table 1.2 Critical depth-of-cuts

Spindle speed (rpm)	Cutting speed (m/min)	Critical DOC (mm)
750	94.7	1.40–1.45
1000	126.3	1.75–1.78
1200	157.8	1.83
1500	189.4	2.21

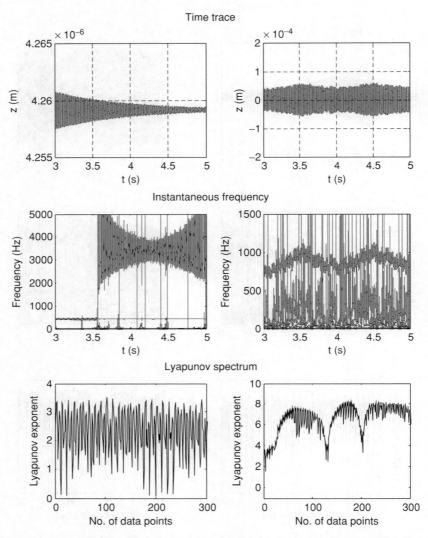

Figure 1.20 Z-direction tool behavior for DOC = 2.21 mm (left) and DOC = 2.22 mm (right) at $\Omega = 1500$ rpm

Figure 1.21 illustrates the comparison of critical DOCs with experimental results subject to increasing speeds. It should be mentioned that the experimental data were taken for a short workpiece and thus the stiffness of the workpiece was higher than the modeled workpiece considered in the section. It was established in previous sections that workpiece dimensions and tool geometry affect cutting stability. However, the workpiece dimensions and tool geometry used for the study are different from the corresponding experimental setup in [7]. Still, the feed rate employed herein differs only by about 5% from those experiments. These differences would certainly contribute to the differences in stability limits.

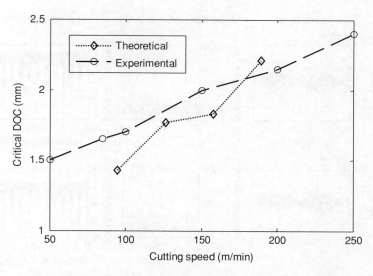

Figure 1.21 Comparison of critical DOC with experimental data

It should be noted that the experimental results in [7] are different from the conventional stability lobes for the ranges of speed that are considered here. Contrary to one's intuition, experiments have proved that critical DOC increases with cutting speed. The machining model also suggests the same.

1.4 Stability in Fine Cuts

Cutting stability subject to DOCs less than 1.00 mm is explored in this section. Three different DOCs at 0.50 mm, 0.75 mm, and 0.90 mm and four different spindle speeds at 750 rpm, 1000 rpm, 1250 rpm, and 1500 rpm are considered. The feed is 0.0965 mm for all cases. All figures have three columns for time traces, instantaneous frequencies, and Lyapunov spectrum. The four rows in the figures represent the results corresponding to four different speeds at a particular DOC.

The X-direction workpiece vibrations for fine cutting are considered below. Figure 1.22 shows the behavior of the case with DOC = 0.50 mm at four different speeds. Time traces for all the four speeds seem similar and their corresponding Lyapunov spectra all demonstrate instability with positive exponents. The workpiece is excited by the tool natural frequency in all cases, except for the case at $\Omega = 1250$ rpm.

When DOC is increased to 0.75 mm, the workpiece becomes more stable in Figure 1.23. Similar to the case with DOC = 0.50 mm, increasing spindle speed does not significantly alter the stability state of the workpiece for DOC = 0.75 mm. All instantaneous frequency plots have a broad bandwidth component at 850–900 Hz. This might be explained by the frequency-doubling of the tool natural frequency. In addition, the Lyapunov spectra do not indicate clean-cut stability with zero exponents. But rather the exponents oscillate near zero

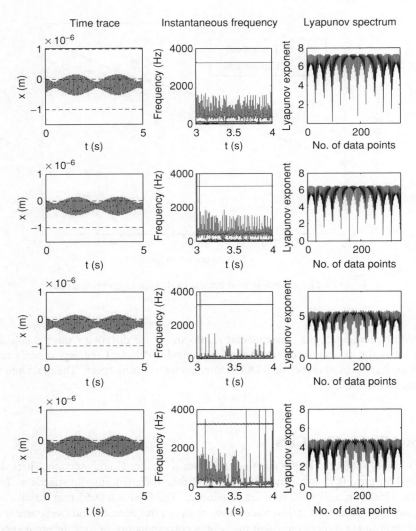

Figure 1.22   X-direction workpiece behavior for DOC = 0.50 mm at (a) $\Omega = 750$ rpm (1st row), (b) $\Omega = 1000$ rpm (2nd row), (c) $\Omega = 1250$ rpm (3rd row), and (d) $\Omega = 1500$ rpm (4th row)

and towards the negative side of the spectra. This can be considered a marginal stable–unstable situation of the workpiece.

Figure 1.24 demonstrates the workpiece behavior subject to DOC = 0.90 mm. It can be seen that the workpiece is in a stable state of motion at this DOC. All the four instantaneous frequency plots illustrate similar types of bifurcated situation having whirling frequency and components other than the workpiece natural frequency. In all cases shown in Figure 1.24, one frequency is between 450–600 Hz and the other extra component is in the range of 1300–1400 Hz. These frequency components do not vary with time or display broadband behaviors.

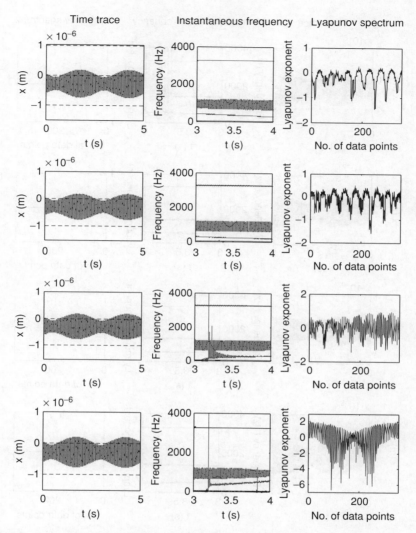

Figure 1.23 X-direction workpiece behavior for DOC = 0.75 mm at (a) $\Omega = 750$ rpm (1st row), (b) $\Omega = 1000$ rpm (2nd row), (c) $\Omega = 1250$ rpm (3rd row), and (d) $\Omega = 1500$ rpm (4th row)

The motion is periodic and stable, as is verified by the zero Lyapunov spectra. It is interesting to note that regardless of the speed in the range considered, the workpiece is dynamically unstable at DOC = 0.50 mm. But it is marginally stable at DOC = 0.75 mm and stable at DOC = 0.90 mm.

The tool stability in the Z-direction investigated in this section considers both time and time-frequency analysis as in the previous sections. Figure 1.25 shows the tool behavior at DOC = 0.50 mm. It is noted that except for the 1250 rpm case, all time traces look similar. In addition, the corresponding instantaneous frequency plots also show the same similarity having bifurcations only in the low frequency region without indicating any broadband behavior.

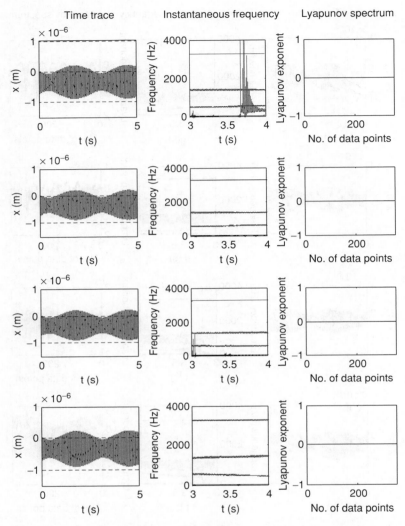

Figure 1.24 X-direction workpiece behavior for DOC = 0.90 mm at (a) $\Omega = 750$ rpm (1st row), (b) $\Omega = 1000$ rpm (2nd row), (c) $\Omega = 1250$ rpm (3rd row), and (d) $\Omega = 1500$ rpm (4th row)

When spindle speed is 1250 rpm in Figure 1.25(c), the tool is excited by the workpiece natural frequency and displays broadband characteristics. Moreover, the associated Lyapunov spectrum indicates an unstable condition, while in all other cases the tool motion is stable with zero exponents.

When DOC is increased to 0.75 mm, all speed cases display similar characteristics having very small vibration amplitudes and bifurcations in the instantaneous frequency plots in Figure 1.26. However, all Lyapunov spectra demonstrate stable behaviors with zero exponents. Increasing DOC further to 0.90 mm, two categories out of the four different speeds in

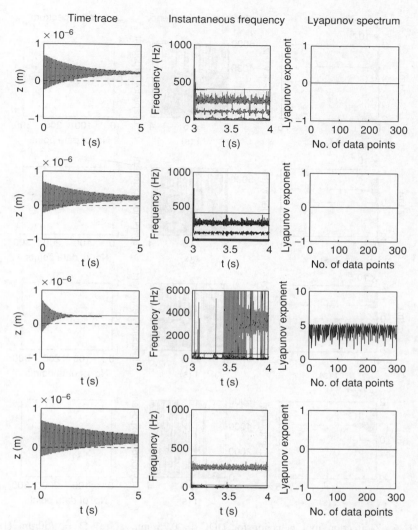

Figure 1.25 Z-direction tool behavior for DOC = 0.50 mm at (a) Ω = 750 rpm (1st row), (b) Ω = 1000 rpm (2nd row), (c) Ω = 1250 rpm (3rd row), and (d) Ω = 1500 rpm (4th row)

Figure 1.27 are seen. The 750 rpm and 1250 rpm cases are alike, while the 1000 rpm and 1500 rpm cases are comparable. Lyapunov spectra show that the 750 rpm and 1250 rpm cases are unstable, and the other two speed cases are stable. Many frequency components are also present in the 1000 rpm and 1500 rpm cases, despite the fact that the system is stable according to the Lyapunov spectrums in Figure 1.28.

Overall system stability can now be determined by examining the stable and unstable situations of the workpiece and the tool. The system is stable if both workpiece and tool demonstrate stable situations. If either of them displays instability, the system is unstable.

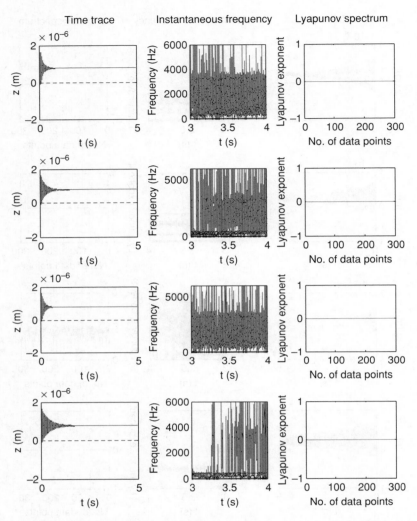

Figure 1.26 Z-direction tool behavior for DOC = 0.75 mm at (a) Ω = 750rpm (1st row), (b) Ω = 1000 rpm (2nd row), (c) Ω = 1250 rpm (3rd row), and (d) Ω = 1500 rpm (4th row)

Using this classification, system stability can be summarized as in Table 1.3 below. Here the notions US, S, and MS stand for Unstable, Stable, and Marginally Stable, respectively. Since the workpiece is unstable for all the speeds at DOC = 0.50 mm, the system is unstable for the DOC. Moreover, the workpiece demonstrates a stable–unstable situation for all four speeds considered and the tool shows highly bifurcated situations for all the cases when DOC = 0.75 mm, thus making the system marginally stable. On the other hand, even though the workpiece is stable at DOC = 0.90 mm for all four speeds, the tool is unstable for the 750 rpm and 1250 rpm cases. Thus the system is one of instability for these two cases.

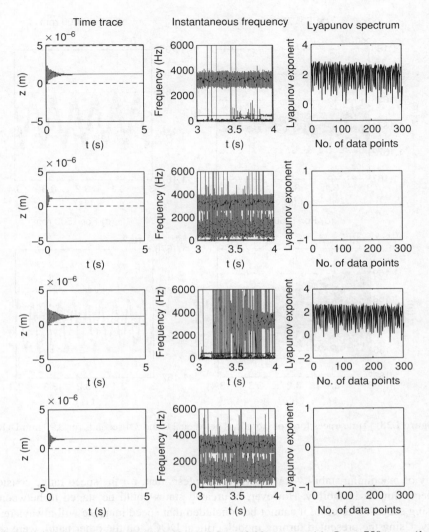

Figure 1.27 Z-direction tool behavior for DOC = 0.90 mm at (a) Ω = 750 rpm (1st row), (b) Ω = 1000 rpm (2nd row), (c) Ω = 1250 rpm (3rd row), and (d) Ω = 1500 rpm (4th row)

Table 1.3 System stability

	Spindle speed (rpm)			
DOC (mm)	750	1000	1200	1500
0.90	US	S	US	S
0.75	MS	US	MS	US
0.50	US	MS	US	MS

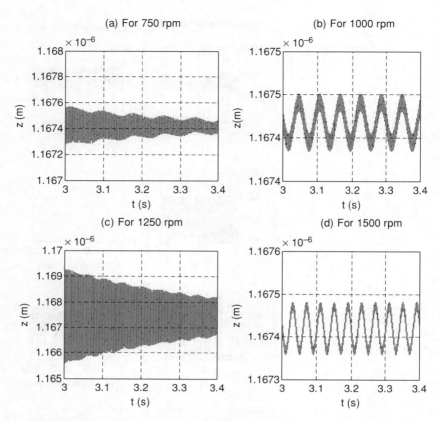

Figure 1.28   Time traces for tool vibrations between 3.0 and 3.4 seconds for 0.90 mm DOC

Study of machining stability for rough cuts suggests that, for the speed range considered, low speeds impart instability. However, bifurcated states could be staged in between with increasing speeds. Therefore it cannot be concluded that speed increase will always result in stability. Using the presented turning model, critical DOCs, on the other hand, were seen to increase with the increase of speed. It was shown that this observation was consistent with experimental data. It was noted that cutting force amplitude oscillations were determined by the nonlinearity of the force, not by speed increments. It was also seen that both the tool and workpiece showed similar instability stages most of the time. However, when it was closer to the critical depth-of-cut, the cutting tool reached instability before the workpiece did. In contrast to this observation, when DOC is less than 1.00 mm, most of the time when the system was unstable, only the workpiece or the tool was unstable and not both. There were a few situations where both became unstable at the same time. It was noted that in all cases where the tool was unstable, the tool was excited by the workpiece natural frequency and displayed broadband behavior. However, not all cases in which the tool was excited by the workpiece natural frequency were unstable. On the other hand, in almost all cases in which the workpiece was unstable, the workpiece had a broadband frequency component near the tool natural frequency. This suggests the importance of considering the workpiece–tool coupling effect in modeling cutting dynamics.

1.5   Concluding Remarks

The three-dimensional, nonlinear dynamic model incorporated *regenerative effect, cutting force nonlinearity, tool nonlinearity, imbalance-induced whirling, and mass and stiffness reduction of the workpiece* to realize simultaneous coupled tool–workpiece vibrations. Model responses subject to whirling were examined for various cases of cutting parameters including depth-of-cut, spindle speed, feed rate, and tool geometry. Stability analysis was performed using instantaneous frequency and Lyapunov spectrum. The following can be summarized based on the results discussed and observations made in the previous sections:

- Concerns over several major issues were raised. It was concluded that workpiece vibrations in modeling turning operation cannot be neglected. The significance of whirling and tool nonlinearity cannot be ignored.
- The various responses of the 3D machining model were shown to agree favorably with experimental data available in the literature.
- Negligence of workpiece vibrations will result in physically inadmissible results.
- Workpiece dimensions impact cutting stability in a non-negligible way.
- Workpiece–tool coupling is significant for the proper interpretation of tool dynamics.
- Use of linearized models for studying cutting dynamics will inevitably generate erroneous, unreliable information.
- Whirling was found to contribute to machining stability and larger workpiece vibration amplitude.

Whirling will affect tool dynamics only when high feed rates are considered. When the critical depth-of-cut is reached and the workpiece vibrates with instability, whirling becomes negligible. Consideration of nonlinearity in modeling machining processes is crucial for understanding the underlying cutting dynamics. Tool geometry significantly affects cutting stability. Machining models disregarding tool geometry will miscomprehend cutting response. Feed rate has an impact on cutting stability. A high feed rate could impart stability to fine turning operation. However, more elaborated investigations are needed to prove that high feed rate does contribute to higher stability limits.

Critical DOC increases with increasing spindle speed. This observation agrees well, in the qualitative sense, with available physical data. Modeling results supported the idea that chatter could occur at low DOCs. This is in agreement with what [8] suggests and what [7] observes in testing. Thus, unlike conventional stability charts, the stability limits established in the chapter have both upper and lower stability regions.

The stability margin for rough cuts using large DOCs can be separated by a line. However, for fine cuts using smaller DOCs there are different stability regions and no stability limits.

Machining chatter can be associated with one of the four types of dynamic stability–instability scenarios: (1) both tool and workpiece are stable, (2) the tool is stable when the workpiece is unstable, (3) the workpiece is stable when the tool is unstable, and (4) both tool and workpiece are unstable.

References

[1] Huo, D., Cheng, K., 2008, "A Dynamics-Driven Approach to The Design of Precision Machine Tools for Micro-Manufacturing and Its Implementation Perspectives," *Proceedings of the Institution of Mechanical Engineers, Part B: Journal of Engineering Manufacture*, 222(1), 1–13.

[2] Dassanayake, A.V., Suh, C.S., 2007, "Machining Dynamics Involving Whirling. Part I: Model Development and Validation," *Journal of Vibration and Control*, 13(5), 475–506.

[3] Dassanayake, A.V., Suh, C.S., 2007, "Machining Dynamics Involving Whirling. Part II: Machining States Described by Nonlinear and Linearized Models," *Journal of Vibration and Control*, 13(5), 507–26.

[4] Dassanayake, A.V., Suh, C.S., 2008, "On Nonlinear Cutting Response and Tool Chatter in Turning Operation," *Communications in Nonlinear Science and Numerical Simulations*, 13(5), 979–1001.

[5] Dassanayake, A.V., 2006, Machining Dynamics and Stability Analysis in Longitudinal Turning Involving Work-piece Whirling, PhD Dissertation, Texas A&M University.

[6] Rao, B. C., Shin, Y. C., 1999, "A Comprehensive Dynamic Cutting Force model for Chatter Prediction in Turning," *International Journal of Machine Tools and Manufacture*, 9(10), 1631–54.

[7] Kim, J.S., Lee, B.H., 1990, "An Analytical Model of Dynamic Cutting Forces in Chatter Vibration," *International Journal of Machine Tools and Manufacture*, 31(3), 371–81.

[8] Volger, M. P., DeVor, R. E., Kapoor, S. G., 2002, "Nonlinear Influence of Effective Lead Angle in Turning Process Stability," *ASME Journal of Manufacturing Science and Engineering*, 124(2), 473–75.

2

Basic Physical Principles

In this chapter the principles upon which the time-frequency control of cutting vibrations is based are briefly reviewed along with the corresponding mathematical equations and some of their important properties. Since time-frequency control employs wavelet transform and adaptive filters as its primary physical components, it is essential that basic topics such as sampling theorem, z-transform, and convolution are discussed. A review of the time-frequency analysis tools commonly adopted for characterizing dynamic instability is also made in the last part of the chapter. These tools are shown to be suitable for linear and stationary signals only. The important conclusion drawn there is that they are not viable for the investigation of turning and cutting operations whose responses are inherently nonlinear and nonstationary. As linear algebra and matrix manipulations are employed to simplify the formulation of the time-frequency control algorithm to be developed in the subsequent chapters, it is beneficial that the mathematical notations used to express the equations are also stated without proof.

2.1 Euclidean Vectors

A Euclidean vector space is a linear functional space that satisfies certain mathematical rules. A vector $\mathbf{V}$ is an ordered series of complex numbers, $v_1, v_2, \ldots, v_n$, defined in the n-dimensional vector space spanned by n mutually orthogonal unit vectors $\{\mathbf{e}_1, \mathbf{e}_2, \ldots, \mathbf{e}_n\}$. $\mathbf{V}$ is a linear combination of its components

$$\mathbf{V} = [v_1, v_2, \ldots, v_n]^{\mathrm{T}} = v_1\,\mathbf{e}_1 + v_2\,\mathbf{e}_2 + \cdots + v_n\,\mathbf{e}_n = \sum_{k=1}^{n} v_k\,\mathbf{e}_k \qquad (2.1)$$

where the superscript T denotes the *transpose* of the vector. The Euclidean inner product of vectors $\mathbf{V}$ and $\mathbf{U}$, which defines the projection of one vector onto the other, is a scalar sum of

Control of Cutting Vibration and Machining Instability: A Time-Frequency Approach for Precision, Micro and Nano Machining, First Edition. C. Steve Suh and Meng-Kun Liu.
© 2013 John Wiley & Sons, Ltd. Published 2013 by John Wiley & Sons, Ltd.

the components of the two vectors

$$\mathbf{V} \cdot \mathbf{U} = \mathbf{V}^{\mathrm{H}} \mathbf{U}$$

$$= \bar{v}_1 u_1 + \bar{v}_2 u_2 + \cdots + \bar{v}_n u_n$$

$$= \sum_{j=1}^{N} \bar{v}_j u_j \tag{2.2}$$

where the superscript H denotes the *Hermitian* of the vector and the overhead bar denotes the complex conjugation of the vector component v_j. $\mathbf{V}$ and $\mathbf{U}$ are orthogonal to each other when $\mathbf{V} \cdot \mathbf{U} = 0$. From the definitions given in Equations (2.1) and (2.2), the projection of $\mathbf{V}$ in the $\mathbf{e}_k$ direction and the projection of $\mathbf{U}$ along the direction of $\mathbf{V}$, $u_{\mathbf{V}}$, can be obtained, respectively, as follows

$$v_k = \mathbf{V} \cdot \mathbf{e}_k \tag{2.3}$$

$$u_{\mathbf{V}} = \mathbf{U} \cdot \left(\frac{\mathbf{V}}{|\mathbf{V}|} \right) \tag{2.4}$$

where $\frac{\mathbf{V}}{|\mathbf{V}|}$ is a unit vector defined by normalizing $\mathbf{V}$ using the vector norm of itself

$$|\mathbf{V}|^2 = \mathbf{V}^{\mathrm{H}} \mathbf{V}$$

$$= \sum_{j=1}^{N} \bar{v}_j v_j$$

$$= \sum_{j=1}^{N} |v_j|^2 \tag{2.5}$$

The corresponding vector norm defined in the three-dimensional Euclidean space is therefore

$$|\mathbf{V}| = \sqrt{v_1^2 + v_2^2 + v_3^2} \tag{2.6}$$

2.2   Linear Spaces

Unlike the Euclidean vector space which is spanned by a set of orthonormal unit vectors, a functional space is in general defined by a group of mutually orthogonal (basis) functions satisfying certain requirements. For example, in the finite energy space $L^2(-\infty, \infty)$, alternatively denoted as $L^2(\mathbf{R})$, the following two conditions have to be satisfied by all functions $f(x)$ in the infinite integral sense

$$\int_{-\infty}^{\infty} |f(x)|^2 \, dx < \infty \tag{2.7}$$

$$\|x\| = \left[\int_{-\infty}^{\infty} |f(x)|^2 \, dx \right]^{1/2} < \infty \tag{2.8}$$

where $\|x\|$ is called the L^2-norm. The norm is commonly used to establish the error between two finite energy functions defined in the root-mean-square sense, such as

$$\|\varepsilon\| = \|f(x) - y(x)\| = \left[\int_{-\infty}^{\infty} |f(x) - y(x)|^2 \, dx \right]^{1/2} \tag{2.9}$$

Error measured in Equation (2.9) can be minimized to eliminate the difference between $f(x)$ and $y(x)$. One such procedure making use of the gradient-based iterative steepest-descent method to minimize $\|\varepsilon\|^2$, instead of $\|\varepsilon\|$, is called the least-mean-square (LMS) algorithm. Adaptive and simple to implement, the LMS algorithm, and its many variations demonstrating various levels of convergence performances, are frequently adopted for *filtering* operation.

The inner product of functions $f(x)$ and $g(x)$ defined in the $\mathbb{C}[a, b]$ space has the following notation

$$\langle f(x), g(x) \rangle = \int_a^b f(t)\overline{g}(t) \, dt \tag{2.10}$$

A function $f(x) \in L^2$ can be represented using a set of orthonormal basis functions $\{\varphi_j(t)\}_{j \in \mathbb{Z}}$ that together span the particular space,

$$f(t) = \sum_{j=-\infty}^{\infty} c_j \varphi_j(t) \tag{2.11}$$

In analogy to Equation (2.3), the coefficients c_j are determined through *orthogonal decomposition* by exploring the property of orthogonality of the basis functions that $\langle \varphi_j, \varphi_k \rangle = 0$, for $j \neq k$ and $\langle \varphi_j, \varphi_k \rangle = 1$, for $j = k$,

$$\begin{aligned}
\langle f, \varphi_k \rangle &= \int_{-\infty}^{\infty} f(t)\overline{\varphi_k(t)} \, dt \\
&= \int_{-\infty}^{\infty} \left(\sum_{j=-\infty}^{\infty} c_j \varphi_j(t) \right) \overline{\varphi_k(t)} \, dt \\
&= c_k
\end{aligned} \tag{2.12}$$

To help elucidate the operation of orthogonal decomposition, let's consider the Fourier transform of a finite time domain function $f(t) \in L^2(\mathbb{R})$, which is defined by the infinite integral

$$\begin{aligned}
F(\omega) &= \int_{-\infty}^{\infty} f(t)e^{-i\omega t} \, dt \\
&= \langle f(t), e^{i\omega t} \rangle
\end{aligned} \tag{2.13}$$

The L^2 normed linear space in which Equation (2.13) is defined has exponential functions $e^{i\omega t}$ as its basis functions that satisfy the orthogonality property. $F(\omega)$, the decomposition coefficient that is also the projection of $f(t)$ on the particular basis function $e^{i\omega t}$, represents the component of $f(t)$ at frequency ω. Once all the components of $f(t)$ are available at all frequencies, $f(t)$ can be reconstructed using the following superposition integral, thus the *inverse* Fourier transform of $F(\omega)$

$$f(t) = \frac{1}{2\pi} \int_{-\infty}^{\infty} F(\omega)e^{i\omega t}\, dt \tag{2.14}$$

2.3   Matrices

An $i \times j$ matrix denoted as $\mathbf{A} = [a_{ij}]$ is an operator in the Euclidean space having a set of entries a_{ij} arranged in a rectangular array as follows

$$\mathbf{A} = [a_{ij}] = \begin{bmatrix} a_{11} & a_{12} & \cdots & a_{1j} \\ a_{21} & a_{22} & \cdots & a_{2j} \\ \vdots & \vdots & \ddots & \vdots \\ a_{i1} & a_{i2} & \cdots & a_{ij} \end{bmatrix} \tag{2.15}$$

If $i = j$, $\mathbf{A}$ is a *square* matrix. In the case that all entries $a_{ij} = 0$, $\mathbf{A}$ is said to be a *null* matrix. If, except for the diagonal entries, all off-diagonal $a_{ij} = 0$ for $i \neq j$, $\mathbf{A}$ is a *diagonal* matrix. An *identify matrix* $\mathbf{I}$ is a special case of a diagonal matrix when $a_{ii} = 1$. The *transpose* of $\mathbf{A}$ is denoted as $\mathbf{A}^T = [a_{ji}]$, thus having the columns and rows interchanged. In the case that $\mathbf{A}^T \mathbf{A} = \mathbf{I}$, $\mathbf{A}$ is an *orthogonal* matrix.

Two matrices $\mathbf{A} = [a_{ij}]$ and $\mathbf{B} = [b_{ij}]$ of the same dimensions may be summed or subtracted to give $\mathbf{C} = \mathbf{A} \pm \mathbf{B} = [c_{ij}]$, with $c_{ij} = a_{ij} \pm b_{ij}$. The product of $\mathbf{A} = [a_{ln}]$ with $\mathbf{B} = [b_{nm}]$, $\mathbf{C} = \mathbf{AB}$, is defined as

$$c_{lm} = \sum_{p=1}^{n} a_{lp} b_{pm} \tag{2.16}$$

The operation in Equation (2.16) is valid only when the number of columns of $\mathbf{A}$ is equal to the number of rows of $\mathbf{B}$. This condition necessarily renders that matrix multiplication is noncommutative; that is, in general, $\mathbf{BA} \neq \mathbf{AB}$.

The *Hermitian transpose* of $\mathbf{A} = [a_{ij}]$, denoted as $\mathbf{A}^H = [\bar{a}_{ji}]$, is the complex conjugate of the transposed matrix. Unitary matrices that satisfy the condition that $\mathbf{A}^H \mathbf{A} = \mathbf{A}\mathbf{A}^H = \mathbf{I}$ also satisfy the important property: $\mathbf{A}^H = \mathbf{A}^{-1}$. If square matrix $\mathbf{A} = \mathbf{A}^H$, then $\mathbf{A} = [a_{ij}]$ is called a *Hermitian* matrix having only real numbers for its diagonal elements a_{ii}. A *Hermitian* $\mathbf{A}$ is *positive-semidefinite* if

$$\mathbf{x}^H \mathbf{A} \mathbf{x} \geq 0 \quad \text{for all } \mathbf{x} \in \mathbb{C}^n \text{ and } \mathbf{x} \neq \mathbf{0} \tag{2.17}$$

The (i, j) *minor* $\mathbf{M}_{ij}$ of a $n \times n$ square matrix $\mathbf{A}$ is the determinant of a submatrix of $\mathbf{A}$ created by removing its ith row and j^{th} column. The (i, j) *cofactor* $\mathbf{C}_{ij}$ of $\mathbf{A}$ is simply

$\mathbf{C}_{i\,j} = (-1)^{i+j}\mathbf{M}_{i\,j}$. The *determinant* of the square matrix $\mathbf{A}$, denoted as $\det(\mathbf{A})$, is defined using its cofactor

$$\det(\mathbf{A}) = \sum_{i,j=1}^{n} a_{ij}\mathbf{C}_{ij} \tag{2.18}$$

The *trace* of $\mathbf{A}$, denoted as $\text{trace}(\mathbf{A})$, is also a scalar

$$\text{trace}(\mathbf{A}) = \sum_{i=1}^{n} a_{ii} \tag{2.19}$$

which is simply the sum of its diagonal entries. Using the cofactor, the *inverse* of the square matrix $\mathbf{A}$, denoted as $\mathbf{A}^{-1}$, that satisfies $\mathbf{A}^{-1}\mathbf{A} = \mathbf{A}\mathbf{A}^{-1} = \mathbf{I}$ is determined by

$$(a_{ij})^{-1} = \frac{\mathbf{C}_{ji}}{\det(\mathbf{A})} \tag{2.20}$$

It should be noted that Equation (2.20) computes the individual entries of the inverse matrix $\mathbf{A}^{-1}$ and that when $\det(\mathbf{A}) = 0$, the matrix is said to be *singular*.

2.3.1　Eigenvalue and Linear Transformation

The *eigenvalues* of an $n \times n$ square matrix $\mathbf{A}$ are the nontrivial solutions λ_i, $i = 1, \ldots, n$ to the following *characteristic equation*

$$\det(\mathbf{A} - \lambda_i \mathbf{I}) = 0 \tag{2.21}$$

which is resulted from considering the following eigenmatrix

$$\mathbf{A}\mathbf{x}_{i \times 1} = \lambda_i \mathbf{x}_{i \times 1} = \mathbf{y}_{i \times 1} \tag{2.22}$$

The physical interpretation of Equation (2.22) is that through the operation of $\mathbf{A}$, the $n \times 1$ vector $\mathbf{x}$ is transformed (through translation or rotation, or both) into vector $\mathbf{y}$ in the same vector space. The components of $\mathbf{x}$ are called *eigenvectors*. To satisfy Equation (2.22) the eigenvectors are necessarily an orthonormal set of linearly independent basis vectors. Some of the important properties of eigenvalues are

$$\text{trace}(\mathbf{A}) = \sum_{i=1}^{n} \lambda_i$$
$$\det(\mathbf{A}) = \prod_{i=1}^{n} \lambda_i \tag{2.23}$$
$$\det(\mathbf{I} + \mathbf{A}) = \prod_{i=1}^{n} (1 + \lambda_i)$$

The eigenmatrix in Equation (2.22) can be recast using a matrix $\mathbf{\Lambda}$ whose diagonal elements carry all the eigenvalues λ_i, $i = 1, \ldots, n$

$$\mathbf{A}\mathbf{x} = \mathbf{x}\mathbf{\Lambda} = \mathbf{x} \begin{bmatrix} \lambda_1 & 0 & \cdots & 0 \\ 0 & \lambda_2 & \cdots & 0 \\ \vdots & \vdots & \ddots & \vdots \\ 0 & 0 & \cdots & \lambda_i \end{bmatrix} \tag{2.24}$$

Since the eigenvectors $\mathbf{x}$ are linearly independent, thus *nonsingular*, we have the following linear transformation that effectively diagonalizes the matrix $\mathbf{A}$ using $\mathbf{x}$

$$\mathbf{x}^{-1}\mathbf{A}\mathbf{x} = \mathbf{\Lambda} \tag{2.25}$$

The interpretation of Equation (2.25) is that by employing the eigenvectors $\mathbf{x}$ as the basis of the vector space, $\mathbf{A}$ is alternatively represented by the diagonal matrix $\mathbf{\Lambda}$. As a linear transformation via the eigenvectors, $\mathbf{x}^{-1}\mathbf{A}\mathbf{x}$ is called a *similarity transformation* on $\mathbf{A}$.

2.4   Discrete Functions

A discrete function defined in the finite-dimensional discrete linear space is a finite set that is at most countable, meaning that it is a group of number sequences such as

$$x(n) = \{x(1), x(2), \ldots, x(N)\}, \ n = 1, 2, \ldots, N \tag{2.26}$$

The rules of operation for the discrete sets in the discrete linear space are similar to those defined for the continuous linear space with a specified norm. The sum and inner product of two discrete sequences are, respectively

$$w(n) = x(n) \pm y(n) = \{x(1) \pm y(1), \ldots, x(N) \pm y(N)\} \tag{2.27}$$

$$\langle x(n), y(n) \rangle = \sum x(n) \, \bar{y}(n) \tag{2.28}$$

The orthogonality decomposition and p-norm are also similarly defined as

$$\langle x(n), \varphi_k(n) \rangle = \sum x(n) \overline{\varphi_k(n)} = c_k \tag{2.29}$$

$$\|x(n)\|_p = \left[\sum_n |x(n)|^p \right]^{1/p} \tag{2.30}$$

Note that when $p = 2$, the norm is the Euclidean norm.

A time function $x(t_n)$ sampled at a constant time interval (sampling period) T with $t_n = n\,\text{T}$ and $n = 0, 1, 2, \ldots, N - 1$, is commonly denoted as $x(n)$. The Fourier transform of the discrete sequence of $x(n)$ is defined using the following notation

$$X(\omega) = \sum_{-\infty}^{\infty} x(n) e^{-in\omega} \tag{2.31}$$

The discrete-time Fourier transform of $x(n)$ in Equation (2.31), is important for understanding filter operation, which is a multiplication process in the Fourier (frequency) domain.

2.4.1   Convolution and Filter Operation

A filter is mathematically a *time-invariant* operator of the specific property that if an input $x(t)$ induces an output $y(t)$ as response, a $\hat{t}$ − time delay of the input $x(t - \hat{t})$ would produce a $\hat{t}$ − time delay of the output $y(t - \hat{t})$ as response. This property also implies *linearity*, meaning that the scaled and summed input $a_1 x_1(t) + a_2 x_2(t)$ induces the scaled and summed response $a_1 y_1(t) + a_2 y_2(t)$.

As a linear operator, a *digital* filter operates on a sequenced input and alters its spectral characteristics as output. The specific operation, called *filtering*, is a *convolution* defining the input–output relationship of the linear time-invariant system which physically is the filter. Filtering $x(n)$ using filter $h(n)$ to generate filtered output $y(n)$ is realized via convolution

$$y(n) = h(n) * x(n) = \sum_k h(k)\, x(n - k) \tag{2.32}$$

The Fourier transform of two convoluted sequences is particularly relevant to filter design. This relevancy is established by the Convolution Theorem which states that the Fourier transform of the convolution of $f(n)$ with $g(n)$ is equivalent to the product of the individual Fourier transform of the two sequences, that is,

$$F\{f(n) * g(n)\} = F\{f(n)\} \cdot F\{g(n)\} \tag{2.33}$$

where $F\{\cdot\}$ is the Fourier transform operator. In other words, in the context of Fourier transform, convolution in the discrete time domain is equivalent to multiplication in the frequency domain. To prove the Convolution Theorem, we express all terms in their explicit forms so that the left-hand side of Equation (2.33) is expanded as follows

$$
\begin{aligned}
F\{f(n) * g(n)\} &= \sum_{-\infty}^{\infty} \left[\sum_k f(k)g(n - k) \right] e^{-in\omega} \\
&= \sum_k f(k) \left[\sum_{-\infty}^{\infty} g(n - k)\, e^{-in\omega} \right]
\end{aligned} \tag{2.34}
$$

The infinite summation term in the brackets can be shown after a simple change-of-variable manipulation to look

$$\sum_{-\infty}^{\infty} g(n - k)\, e^{-in\omega} = e^{-ik\omega}\, F\{g(n)\} \tag{2.35}$$

By substituting the result above we have

$$F\{f(n) * g(n)\} = \sum_k f(k) \left[e^{-ik\omega} F\{g(n)\} \right]$$

$$= F\{g(n)\} \left[\sum_k f(k) e^{-ik\omega} \right]$$

$$= F\{g(n)\} \cdot F\{f(n)\} \tag{2.36}$$

thus completing the proof.

2.4.2 Sampling Theorem

A finite energy time function $x(t)$ is sampled at a constant time interval T with $t_n = n\,$T and $n = 0, 1, 2, \ldots, N - 1$. If the sampling rate, $f_s = {}^1/_T$ is higher than twice the highest frequency of $x(t)$, then the sampled sequence of $x(t)$, denoted as $x(nT)$, contains all the information of the time function. The formal statement of the above, as stated in the following, is the sampling theorem: if $x(t)$ is bandlimited to W Hz frequency, that is, $-2\pi W < F\{x(t)\} < 2\pi W$, then it can be fully represented by the sampled sequence $x(n)$ taken at a sampling rate $f_s > 2\,W$ using

$$x(t) = \sum_{-\infty}^{\infty} x(nT) g(t - nT) \tag{2.37}$$

where

$$g(t) = \mathrm{sinc}(\pi f_s t) = \frac{\sin(\pi f_s t)}{\pi f_s t} \tag{2.38}$$

In other words, $x(t)$ can be reconstructed without error by interpolation in time between samples using the sinc function defined in Equation (2.38).

Since

$$\langle g(t - nT), g(t - mT) \rangle = \int_{-\infty}^{\infty} g(t - nT)\, \bar{g}(t - mT)\, dt$$

$$= \begin{cases} 0, & m \neq n \\ T, & m = n \end{cases} \tag{2.39}$$

then $g(t) = \mathrm{sinc}(\pi f_s t)$ is a set of orthogonal functions. Now consider the following Fourier series

$$\int_{-\infty}^{\infty} x(t)\, \bar{g}(t - mT)\, dt = \sum_{-\infty}^{\infty} x(nT) \int_{-\infty}^{\infty} g(t - nT)\, \bar{g}(t - mT)\, dt$$

$$= x(mT) T \tag{2.40}$$

where in deriving the above Equation (2.37) is used and summation and integration are interchanged. From Equation (2.40), the sampled sequence $x(mT)$ can be obtained as the coefficient of the infinite orthogonal expansion in Equation (2.37):

$$x(mT) = \frac{1}{T} \int_{-\infty}^{\infty} x(t)\bar{g}(t - mT)dt \qquad (2.41)$$

2.4.3   z-Transform

The z-transform is the tool of choice of many for designing digital filter banks. The transform maps a discrete sequence of numbers in the time domain into the z-domain to get a complex representation similar to the discrete Laplace transform:

$$X(z) = \sum_{-\infty}^{\infty} x(n)z^{-n} \qquad (2.42)$$

The analogy of the notion to the Fourier transform is readily noted. By replacing z with $e^{i\omega}$, Equation (2.42) becomes

$$X(z = e^{i\omega}) = \sum_{-\infty}^{\infty} x(n)e^{-in\omega} \qquad (2.43)$$

Thus it is understood that the z-transform allows time sequence $x(n)$ to be manipulated in the complex domain with the underlying spectral information being kept intact.

Of the many properties of the z-transform, the *region of convergence* (ROC) is essential for the determination of system (input–output) stability. ROC is the set of z values that satisfies the convergence condition

$$\left| \sum_{-\infty}^{\infty} x(n)z^{-n} \right| < \infty \qquad (2.44)$$

Once filter stability is established in the complex z-domain, the original time sequence $x(n)$ can be recovered using the *inverse* z-transform

$$x(n) = \frac{1}{2\pi i} \oint_C X(z)z^{n-1}dz \qquad (2.45)$$

where the contour integral is to be performed counterclockwise along the closed path of the ROC. One other important property of the z-transform is similar to Equation (2.33) as follows

$$Z\{f(n) * g(n)\} = Z\{f(n)\} \cdot Z\{g(n)\}$$
$$= F(z) \cdot G(z) \qquad (2.46)$$

with $Z\{\cdot\}$ being the z-transform operator and $F(z)$ and $G(z)$ understood to be the z-transform of $f(n)$ and $g(n)$, respectively. Not surprisingly, convolution in the discrete time domain is equivalent to multiplication in the complex z-domain.

2.5 Tools for Characterizing Dynamic Response

It is essential that a cutting process is thoroughly characterized for nonlinear response and instability before proper control can be applied. The response of a dynamic system in general could be a fixed point, a periodic solution, or a nonperiodic solution. A stable linear or almost linear system would have an equilibrium point or a fixed point as its response. The Hartmen–Grobman Theorem states that the dynamic characteristics of a perturbed system are qualitatively similar to its linear counterpart in the neighborhood of the equilibrium point [1] and that the stability near the equilibrium point can thus be determined by slightly perturbing this point. If the system restores back to the original point, the system near the point is stable. Otherwise, if the system diverges away from the point, it is unstable.

By studying the geometric characteristics and flow paths of the solution trajectory in the state space, the stability near the trajectory can be determined using phase portraits. Powerful graphical tools that aid in identifying the dynamic state of a system also include Poincaré maps and bifurcation diagrams. A Poincaré map is a collection of sequential points in the state-space generated by the penetration of a continuous evolution trajectory through a generalized surface or plane in the space [2]. For a periodically forced, second-order nonlinear oscillator, a Poincaré map can be obtained by stroboscopically observing the position and velocity at a particular phase of the forcing function. A bifurcation diagram is used to plot the transition of a motion from stability to chaos, as some parameters are varied. Such diagrams can be obtained by time-sampling the motion as in a Poincaré map and displaying the output against the control parameter. The bifurcation diagram of the *quadratic map* shown in Figure 2.1 is given as an example, in which f_μ is plotted against the control parameter μ using the iterative scheme defined below

$$f_\mu(x) = \mu x(1 - x) \quad x \in [0, 1]$$
$$y = f_\mu^n(x) = f_\mu(f_\mu(f_\mu \ldots (x))) \tag{2.47}$$

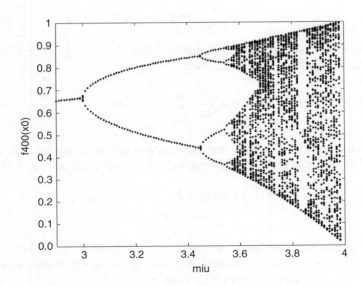

Figure 2.1 Bifurcation diagram of the quadratic map

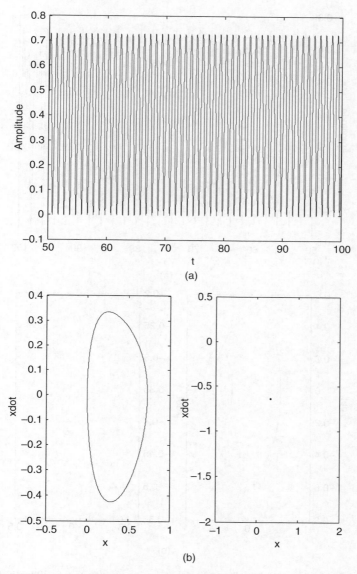

Figure 2.2 (a) Time history of a nonlinear model at driving frequency $\omega = 100\pi$ (b) Corresponding phase portrait and Poincaré map of the response in (a)

It can be seen that period-doubling bifurcation occurs at $\mu = 3$ and another bifurcation takes place at approximately $\mu = 3.47$. Through a cascade of period-doubling bifurcation, the map becomes chaotic at $\mu = 4$.

For a trajectory that follows a close orbit in the phase portrait and returns precisely to where it first started after one period, T, the motion is periodic and the closed orbit is called a limit cycle. The trajectory of a periodic motion, starting initially either close to or sufficiently

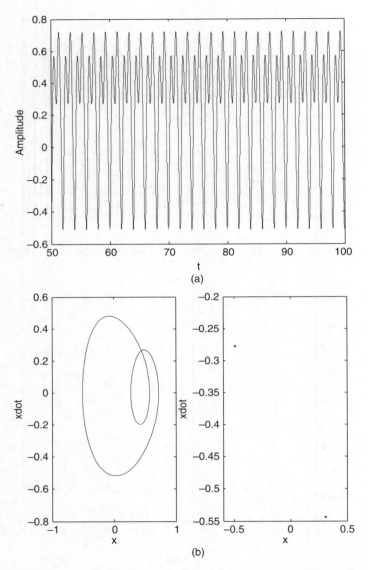

Figure 2.3   (a) Time history of a nonlinear model at driving frequency $\omega = 200\pi$ (b) Corresponding phase portrait and Poincaré map of the response in (a)

away from the origin, always converges to a limit cycle in the phase space when this periodic solution is stable. Periodic solution will locate at the same position in the corresponding Poincaré map. The Poincaré map is a qualitative topological approach widely applied to the prediction of chaos and study of stability in the state space through exploring the geometric features of the sequence of points on the Poincaré section. This Poincaré mapping has the same general stability as the flow in the state space [3]. When a system undergoes chaotic motion, its associated Poincaré map will show specific shapes and features indicating the state and extent of bifurcation. Figure 2.2 shows the time history, phase portrait and Poincaré map of a nonlinear response in [4]. Figure 2.2(b) shows that there is one close orbit in the phase portrait and one point in the Poincaré map.

Consider the autonomous system defined below

$$\dot{\mathbf{x}} = \mathbf{f}(\mathbf{x}, \lambda) \tag{2.48}$$

Let $\mathbf{x}_0(t)$, function of a minimal period T, be the periodic solution to Equation (2.48). Impose a perturbation $\mathbf{y}$ on $\mathbf{x}_0(t)$

$$\mathbf{x}(t) = \mathbf{x}_0(t) + \mathbf{y}(t) \tag{2.49}$$

and substitute it into Equation (2.48), the Taylor-series expansion of the resulted equation about $\mathbf{x}_0(t)$ is thus

$$\dot{\mathbf{y}}(t) = D_x\mathbf{F}(\mathbf{x}_0(t), \lambda)\mathbf{y}(t) + O(\|\mathbf{y}(t)\|) \tag{2.50}$$

When n perturbations are applied to $\mathbf{x}_0(t)$, the n-dimensional linear system represented by Equation (2.50) has n linearly independent solutions, $\mathbf{y}_i(t)$. These n solutions can be expressed as a $n \times n$ matrix

$$\mathbf{Y}(t) = [\mathbf{y}_1(t), \mathbf{y}_2(t), \ldots, \mathbf{y}_n(t)] \tag{2.51}$$

The following relation satisfies Equation (2.50)

$$\dot{\mathbf{Y}} = D_x\mathbf{F}(\mathbf{x}_0(t), \lambda)\mathbf{Y} \tag{2.52}$$

Since $\mathbf{x}_0(t) = \mathbf{x}_0(t + T)$, it can be derived that

$$D_x\mathbf{F}(\mathbf{x}_0(t), \lambda) = D_x\mathbf{F}(\mathbf{x}_0(t + T), \lambda) \tag{2.53}$$

If $\mathbf{Y}(t) = [\mathbf{y}_1(t), \mathbf{y}_2(t), \ldots, \mathbf{y}_n(t)]$, then $\mathbf{Y}(t) = [\mathbf{y}_1(t + T), \mathbf{y}_2(t + T), \ldots, \mathbf{y}_n(t + T)]$. Because Equation (2.50) has at most n linearly independent solutions, and $\mathbf{Y}(t)$ are such n linearly independent solutions, $\mathbf{Y}(t + T)$ must be the linear combination of $\mathbf{y}_1(t), \mathbf{y}_2(t), \ldots, \mathbf{y}_n(t)$, thus meaning

$$\mathbf{Y}(t + T) = \mathbf{Y}(t)\Phi \tag{2.54}$$

The matrix Φ is a transformation from an n-dimensional vector at $t = 0$ to another vector at $t = T$. Assuming an initial condition $\mathbf{Y}(0) = \mathbf{I}$, where $\mathbf{I}$ is an identity matrix, and $t = 0$ in Equation (2.54), we have

$$\varphi = \mathbf{Y}(T) \tag{2.55}$$

The matrix $\mathbf{\Phi}$ is called *monodromy matrix*. The reason for $\mathbf{Y}(0) = \mathbf{I}$ is because the differentiation of the solution at $t = 0$ with respect to the initial condition $\mathbf{x}_0$ is an identity matrix. Differentiating Equation (2.48), we have

$$\frac{d}{dt}\left(\frac{\partial \mathbf{x}}{\partial \mathbf{x}_0}\right) = D_x \mathbf{F}(\mathbf{x}(t), \lambda) \frac{\partial \mathbf{x}}{\partial \mathbf{x}_0} \tag{2.56}$$

And differentiating the initial condition $\mathbf{x}(0) = \mathbf{x}_0$, with respect to $\mathbf{x}_0$, we have

$$\frac{\partial \mathbf{x}(0)}{\partial \mathbf{x}_0} = \mathbf{I} \tag{2.57}$$

Equation (2.57) is an initial value problem with Equation (2.57) as its initial condition that $\frac{\partial \mathbf{x}(0)}{\partial \mathbf{x}_0} = \mathbf{Y}(0) = \mathbf{I}$ and $\frac{\partial \mathbf{x}(T)}{\partial \mathbf{x}_0} = \mathbf{Y}(T) = \mathbf{\Phi}$ is the monodromy matrix. Differentiating Equation (2.48) with respect to t,

$$\ddot{\mathbf{x}} = D_x \mathbf{F}(\mathbf{x}(t), \lambda)\dot{\mathbf{x}} \tag{2.58}$$

If $\mathbf{x}(t)$ is a solution of Equation (2.48), then $\dot{\mathbf{x}}(t)$ is a solution of Equation (2.58), thus also of Equation (2.50). In addition, because

$$\mathbf{x}_0(t) = \mathbf{x}_0(t + T) \Rightarrow \dot{\mathbf{x}}_0(t) = \dot{\mathbf{x}}_0(t + T) \tag{2.59}$$

when $t = 0$, we have

$$\dot{\mathbf{x}}(0) = \dot{\mathbf{x}}_0(T) \tag{2.60}$$

Because $\dot{\mathbf{x}}_0(t)$ is a solution of Equation (2.50), it is therefore a linear combination of $\mathbf{y}_1(t), \mathbf{y}_2(t), \ldots, \mathbf{y}_n(t)$, thus of the form

$$\dot{\mathbf{x}}_0(t) = \mathbf{Y}(t)\,\mathbf{c} \tag{2.61}$$

where $\mathbf{c}$ is a constant scalar vector. Evaluating Equation (2.61) at $t = 0$ and $t = T$, we obtain

$$\dot{\mathbf{x}}_0(0) = \mathbf{Y}(0)\mathbf{c} \quad \text{and} \quad \dot{\mathbf{x}}_0(T) = \mathbf{Y}(T)\mathbf{c} \tag{2.62}$$

Equation (2.60) implies the following

$$\mathbf{Y}(0)\mathbf{c} = \mathbf{Y}(T)\mathbf{c} \tag{2.63}$$

From $\mathbf{Y}(0) = \mathbf{I}$ and also Equation (2.55), Equation (2.63) can be written as

$$\mathbf{\Phi}\,\mathbf{c} = \mathbf{c} \tag{2.64}$$

thus showing that 1 is an eigenvalue of the matrix $\mathbf{\Phi}$ corresponding to the eigenvector $\mathbf{c} = \mathbf{x}_0(0)$. The stability of periodic solutions can be determined by other eigenvalues of $\mathbf{\Phi}$ called the

Floquet multipliers. According to the Floquet theory, if the magnitudes of these eigenvalues are less than 1, the periodic solution is stable. If one or more than one of their magnitude are greater than 1, the periodic solution is not stable and bifurcations would occur. The type of bifurcation can be determined by the way the eigenvalue(s) behaves outside of the unit circle.

A periodic motion may become unstable if the control parameters are allowed to vary – a scenario signifying dynamic deterioration of stability that could lead to eventual chaos. Period-doubling, secondary Hopf bifurcation, intermittence, and crises are several possible routes-to-chaos. Period-doubling is a commonly observed route through which a machining system becomes chaotic [4, 5, 6]. When the control parameter slowly varies, period-doubling bifurcation will occur when one of the Floquet multipliers goes out of the unit circle from -1. This is depicted in the Poincaré section as two points. In addition, this frequency-halving bifurcation would occur at smaller and smaller intervals of the control parameter and chaos would occur beyond a critical value as shown in Figure 2.1. Figure 2.3 shows the time waveform, phase portrait, and Poincaré map of a nonlinear rotor-dynamic response in [7]. Figure 2.3(b) shows that there are two closed orbits and there are two points in its corresponding Poincaré map.

A *quasi-periodic* motion is a dynamic solution characterized by two or more incommensurate frequency components; that is, when the ratio of two frequencies is an irrational number. Quasi-periodic motions come from the secondary Hopf bifurcation. In terms of Floquet multipliers, the Hopf bifurcation is marked by having two complex-conjugate multipliers crossing the unit circle simultaneously. Again, if the control parameter were further varied, the motion would become chaotic. Appearance of quasi-periodic motion indicates the set-in of dynamics instability. Although the presence of many incommensurate frequencies in the motion inevitably complicates the time waveform and phase portrait of a quasi-periodic motion, this kind of response can be determined using Poincaré maps and Fourier transform. The corresponding phase portrait will show complex closed orbits and the Poincaré map will contain one or more closed-orbit-like point groups. Figure 2.4 shows the waveform, phase portrait, and Poincaré map of a quasi-periodic response. This response is obtained from a time-delayed model in [8]. It can be seen in Figure 2.4(b) that the phase plot of the quasi-periodic response is heavily intertwined and its Poincaré map shows a closed point sequence.

These graphical tools are powerful for providing qualitative measures and therefore are not feasible for quantitatively establishing whether an erratic response is indeed chaotic or not. As a quantitative measure, the Lyapunov exponents and fractal dimensions were developed to determine if a motion is in the state of full-fledged or weak chaos. Some systems could eventually display certain *fractal* features in their Poincaré section subject to continual dynamic deterioration. A fractal state is the transition from a quasi-periodic motion or period-doubling to a chaotic motion. A fractal or fractal set has fine details at all possible scales and the quality of self-similarity at different scales; however, there is not a universal definition for fractal. Fractal dimension is a quantitative property of a set of points in an n-dimensional space that measures the extent to which the points fill a sub-space, as the number of points becomes very large [9]. The fractal dimension of an infinite fractal set of points is generally noninteger and less than n. The phase portrait of a fractal response is characterized by many tightly linked and hard to discern closed orbits. Poincaré maps are generally considered a better alternative to the phase portrait for revealing the physical essence of a fractal response. Figure 2.5 shows the phase portrait and Poincaré map of a fractal response acquired from [4], in which crack-induced rotor-dynamic nonlinearities were investigated. The fractal structure is clearly seen.

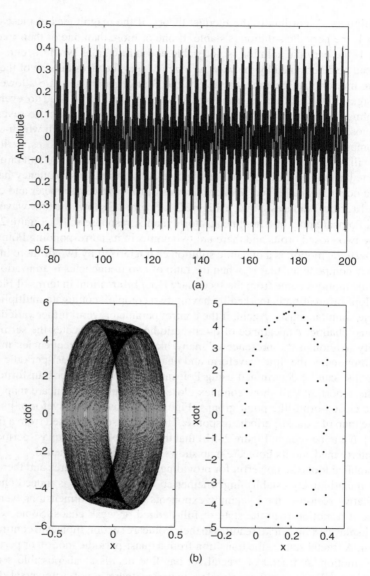

Figure 2.4 (a) Time history of a nonlinear time-delayed model (b) Corresponding phase portrait and Poincaré map of the response in (a)

Because they are both nonperiodic and spectrally broadband, it is difficult to distinguish a fractal motion from a chaotic motion. Chaos is irregular and unpredictable, and can be thought of as a bounded nonperiodic behavior whose attractors are geometrical objects that possess fractal dimensions. Because of its relatively broadband spectrum, characterized by spikes indicating the dominant frequencies, a chaotic response can also be considered the superposition of a very large number of unstable motions. Chaotic systems are also characterized by their

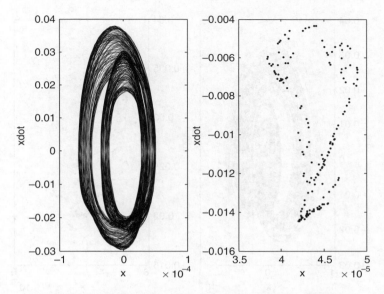

Figure 2.5   Phase portrait and Poincaré map of a fractal response

sensitivity to perturbations to the initial condition; that is, a tiny variation in the input can be quickly amplified to create overwhelmingly disproportionate output. As expected, it is difficult to make out a chaotic motion from among many fractal-likes using the phase portrait. However, as chaos can sometimes be readily identified with the forming of certain geometric shapes, such as a horseshoe or a spoon in the Poincaré space, the Poincaré map is one of the alternatives for positively identifying a chaotic response. Figure 2.6 shows the phase portrait and Poincaré map of a chaotic response also from [4], where a strange geometric shape appears in the Poincaré map.

A powerful tool for the same objective, the Lyapunov exponent is a quantitative measure of the exponential attraction or repulsion in time of two adjacent trajectories in the state space with different initial conditions. Chaos can be distinguished from noisy behavior due to random external influences, and its degree evaluated using Lyapunov exponents. A positive Lyapunov exponent would thus indicate a chaotic motion with a bounded trajectory. However, as different attractors can have the same exponents or dimensions, descriptions of chaotic response using the Lyapunov exponents or fractal dimensions are often insufficient [10]. As such, these time domain methods, which are initial-condition dependent and also require that all data be available for analysis, rely on the expertise and subjective discretion of an experienced person for sound judgment calls.

2.5.1   Fourier Analysis

Bifurcation is the transition of a motion from one state to another accompanied by the appearance of new modes, the disappearance of old modes, or both. Thus, different states of periodic, period-doubling, quasi-periodic, and chaotic motions can be readily identified by their

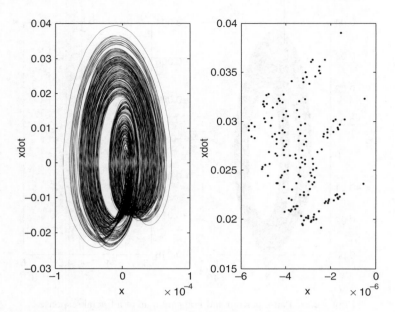

Figure 2.6   Phase portrait and Poincaré map of a chaotic response

respective spectrum. This is one of the reasons that Fourier-based methods are often used for the identification of bifurcations. The assumption made in the method is that a periodic or nonperiodic signal can be represented as a synthesis of sine or cosine functions

$$x(t) = \frac{1}{2\pi} \int_{-\infty}^{\infty} F(\omega)e^{i\omega t}d\omega = \sum_{j=1}^{\infty} a_j e^{i\omega_j t} \qquad (2.65)$$

where $e^{i\omega t} = \cos(\omega t) + i\sin(\omega t)$.

Since the transform $F(\omega)$ is often complex, the absolute value $|F(\omega)|$ is used in graphical displays. When the motion is periodic or quasi-periodic, $|F(\omega)|$ shows a set of narrow spikes or lines indicating that the signal can be represented by a discrete set of harmonic functions. Near the onset of chaos, however, a continuous distribution appears, and in the fully chaotic regime the continuous spectrum may dominate the discrete spikes. Figure 2.7 shows the spectra of the periodic motion given in Figure 2.2, the period-doubling response in Figure 2.3, the quasi-periodic motion in Figure 2.4, and the chaotic motion in [11] (for a milling data at 2700 rpm spindle speed). In Figure 2.7(a), there is only one frequency component for the periodic solution. There is a frequency-halving phenomenon for the period-doubling bifurcation in Figure 2.7(b). There are two incommensurate frequencies in the spectrum of the quasi-periodic motion in Figure 2.7(c). The chaotic motion has a broadband spectrum as shown in Figure 2.7(d) in which continuous frequency distribution is seen occurring.

However, the Fourier transform in Equation (2.65) uses a set of linear periodic harmonic functions with constant amplitude and constant frequency to represent nonlinear and non-periodic signals, which is clearly against our understanding about nonlinear nonstationary

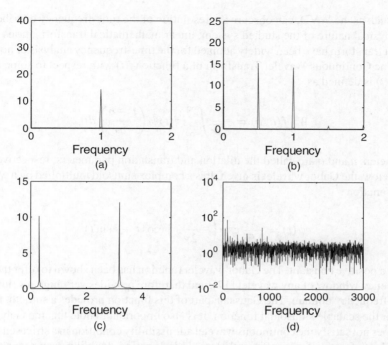

Figure 2.7 The spectrum of (a) a periodic response, (b) a period-doubling response, (c) a quasi-periodic response, and (d) a chaotic motion

signals. For the obvious reasons that the Fourier transform is linear and that its analyzing harmonic functions are stationary, spectral domain methods are not feasible for time-varying series demonstrating nonlinearity [12]. What, then, are the fundamental attributes required of a viable approach effective for characterizing nonlinear states of motion and for identifying bifurcations as they occur and as they are in transition? Because this frequency domain method totally loses the time domain information, which is essential for the detection of the occurrence of bifurcation and state change in *real-time*, many research efforts have been dedicated to developing an approach that can simultaneously detect small variations in both the time and frequency domains.

2.5.2   Wavelet Analysis

A good way of detecting and identifying changes in dynamic states and bifurcations is to monitor spectral variations in time, thus implying characterizing a signal in both the time and frequency domains. The objective of time-frequency analysis is to describe how the spectral content of a signal evolves and to establish the physical and mathematical essences needed for understanding a time-varying spectrum. Among the various time-frequency analysis methods applied to dynamical nonlinear analyses and fault detection [4, 13, 14] are the Short-Time Fourier Transform (STFT), wavelet analysis, Wigner–Ville Distribution (WVD), Choi–Williams Methods, and Born–Jordan Distribution [15, 16]. Since the cross-items of bilinear methods, such as the WVD, can obscure the resolution of the time-frequency distribution, and thus the physical nature of the studied system, linear mathematical transformations including the wavelet transform have been widely adopted for the time-frequency analysis of nonlinearity [15, 16]. The Continuous Wavelet Transform of a function $f(t)$ with respect to some analyzing wavelet $\psi(t)$ is defined as

$$W_\psi f(a, b) = \frac{1}{\sqrt{a}} \int_{-\infty}^{\infty} f(t)\bar{\psi}\left(\frac{t-b}{a}\right) dt \qquad (2.66)$$

The parameters a and b are called the dilation and translation parameters, respectively. In the equation below, the Gabor wavelet is given. It is a complex sinusoid multiplied by a windowing Gaussian function

$$\psi(t) = \frac{1}{\sqrt{\gamma}} \exp\left(-\frac{t^2}{2\gamma^2}\right)(\cos t + i \sin t) \qquad (2.67)$$

where γ is a positive constant. The Gabor wavelet function has been shown to offer the smallest time-frequency window of any wavelet [17], and therefore provides very high resolution in the joint time-frequency domain. The Gaussian part of this function provides a smooth, finite time window for the scalable sinusoidal function. It is also important to note that the Gabor wavelet function does not satisfy the continuous wavelet admissibility condition in a strict sense, but can be assumed satisfactory if γ is sufficiently large [18]. If period-doubling occurred in a dynamic system, a new frequency at half of the forcing frequency would appear in the time-frequency domain. If the system continues to deteriorate, more period-doublings would occur and more low-order subharmonic frequency components would appear in the time-frequency domain. Similar situations would also arise for the secondary Hopf bifurcation, where a frequency component incommensurate to the forcing frequency would appear afterwards. Figure 2.8 shows the period-doubling case, again, from [19], in which case the forcing frequency is

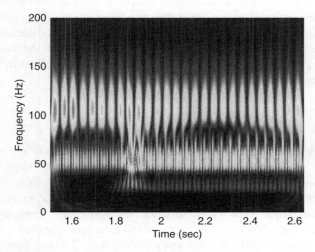

Figure 2.8   Gabor wavelet transform of a signal undergoing period-doubling

50 Hz and the 25 Hz frequency component corresponding to the period-doubling bifurcation appears at 1.8 sec.

Even though time-frequency distribution does work for some simple cases, it cannot provide the necessary resolution for analyzing nonlinear, nonstationary signals and identifying bifurcation and system dynamic states. To localize an event in time, the time window width has to be narrow, thus inevitably resulting in the dilemma of poor frequency resolution. The dilemma is inherent in all Fourier-based time-frequency distributions. Although their time windows can be adjusted, nevertheless wavelets in general provide uniformly unsatisfactory resolution [12]. All Fourier-base time-frequency methods are of poor time resolution in low frequency. Since all time events within the time window are averaged over the entire window, for example, Fourier method cannot distinguish two pulses from each other [20]. To define a local change using wavelets, since the higher the frequency the more localized the basic wavelet will have to be, one must find the result in the high frequency range. Even if a local event happens only in the low frequency range, one would still be forced to find its effects in the high frequency range, thus making the task of interpreting nonlinear, nonstationary signals more difficult. In addition, as they are all Fourier-based and thus effective for analyzing linear stationary signals, the abovementioned time-frequency methods are unsuitable for nonlinear, nonstationary signal analysis. Using the example in Figure 2.8, should the bifurcation occur in the frequency range higher than 150 Hz, the newborn frequency would be obscured by the forcing frequency and other frequencies, thus falsifying the detection and identification of the bifurcation. Due to its low time-frequency resolution in resolving chaotic responses that are characteristically nonperiodic and broadband, and in locating the time instant at which bifurcation occurs, the result is at best nonconclusive for this case. Figure 2.9 shows the time-frequency distribution of a milling chatter response obtained at 2700 rpm spindle speed [11]. Since all the modes are mixed together due to the averaging effect inherent in the Fourier-based time window, only the forcing frequency is unambiguously visible. No other frequency can be discerned and no bifurcation can be detected or identified.

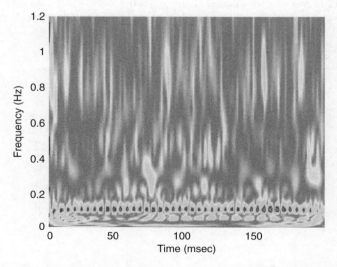

Figure 2.9 Gabor wavelet transform of a signal experiencing milling chatter

References

[1] Nayfeh, A.H., Balachandran, B., 1994, *Applied Nonlinear Dynamics*, John Wiley and Sons, New York.

[2] Moon, F.C., 1987, *Chaotic Vibration: An Introduction for Applied Scientists and Engineers*, John Wiley and Sons, New York.

[3] Thompson, J.M.T., Stewart, H.B., 2001, *Nonlinear Dynamics and Chaos*, John Wiley and Sons, New York.

[4] Yang, B., Suh, C. S., Chan, A. K., 2002, "Characterization and Detection of Crack-Induced Rotary Instability," *ASME Journal of Vibration and Acoustics*, 124(1), 40–48.

[5] Liu, M.-K., Suh, C.S., 2012, "On Controlling Milling Instability and Chatter at High-Speed," *Journal of Applied Nonlinear Dynamics*, 1(1), 59–72.

[6] Dassanayake, A.V., Suh, C.S., 2008, "On Nonlinear Cutting Response and Tool Chatter in Turning Operation," *Communications in Nonlinear Science and Numerical Simulations*, 13(5), 979–1001.

[7] Yang, B., 2003, On The Characterization of Fault-Induced Rotor-Dynamic Bifurcations and Nonlinear Responses, Ph.D. Dissertation, Texas A&M University, College Station, TX.

[8] Yang, B., Suh, C. S., 2004, "On the Nonlinear Features of Time-delayed Feedback Oscillators," *Communications in Nonlinear Science and Numerical Simulations*, 9(5), 515–29.

[9] Moon, F. C., 1992, *Chaotic and Fractal Dynamics*, John Wiley and Sons, New York.

[10] Seydel, R., 1994, *Practical Bifurcation and Stability Analysis*, Springer-Verlag, New York.

[11] Suh, C. S., Khurjekar, P. P., Yang, B., 2002, "Characterization and Identification of Dynamic Instability in Milling Operation," *Mechanical Systems and Signal Processing*, 15(5), 829–48.

[12] Huang, N.E., Shen, Z., Long, S. R., *et al.*, 1998, "The Empirical Mode Decomposition and Hilbert Spectrum for Nonlinear and Nonstationary Time Series Analysis," *Proceedings of Royal Society, London Series A*, 454(1971), 903–95.

[13] Newland, N.E., 1994, "Wavelet analysis of Vibration: Theory and Wavelet Maps," *ASME Journal of Vibration and Acoustics*, 116(4), 409–25.

[14] Gaberson, H. A., 2002, "The Use of Wavelets for Analyzing Transient Machinery Vibration," *Sound and Vibration*, September 2002, 12–17.

[15] Cohen, L., 1995, *Time-Frequency Analysis*, Prentice Hall PTR, Upper Saddle River, New Jersey.

[16] Qian, S., Chen, D., 1996, *Joint Time-Frequency Analysis*, Prentice Hall PTR, Upper Saddle River, New Jersey.

[17] Chui, C.K., 1992, *An Introduction to Wavelets*, Academic Press, San Diego, CA.

[18] Kishmoto, K., Inoue, H., Hamada, M., Shibuya, T., 1995, "Time Frequency Analysis of Dispersive Wave by Means of Wavelet Transform," *ASME Journal of Applied Mechanics*, 62(4), 841–46.

[19] Yang, B., Suh, C. S., 2004, "On the Characteristics and Interpretation of Bifurcation and Nonlinear Dynamic Response," *ASME Journal of Vibration and Acoustics*, 126(4), 574–79.

[20] Rioul, O., Vetterli, M., 1991, "Wavelet and Signal Processing," *IEEE Signal Processing Magazine*, 8(1), 14–38.

3

Adaptive Filters and Filtered-x LMS Algorithm

Digital filtering as defined in Equation (2.32) is a linear operation that modifies the input sequence $x(n)$ using the time-invariant filter $h(n)$ to generate an output sequence $y(n)$ of certain desired properties. When the filter coefficients that define the time domain magnitude and spectral responses of $h(n)$ are allowed to vary with time and be adjusted according to a set of computational rules, the filter is termed designable and adaptive. Adaptive filters are commonly adopted for anticipating situations when the required spectral response of the modification operation is not available *a priori*. They are notably applicable to pre-conditioning inputs that are stochastic and nonstationary. This particular applicability has significant implications for the control of nonlinear dynamic systems whose responses are characteristically aperiodic and time varying. The basic features of adaptive filtering relevant to time-frequency control of vibration are reviewed in the chapter. To establish the working principle behind the Filtered-x Least-Mean-Square (FXLMS) control algorithm, which is an essential component of the time-frequency control, it is necessary that the concepts of minimum mean-square error criterion, Wiener filters, and LMS adaptive algorithm are introduced first.

3.1 Discrete-Time FIR Wiener Filter

The optimal Wiener filter is the direct result of solving the Wiener–Hopf integral equations using the concept of adaptive filters [1]. Consider the block schematic of an FIR adaptive filter in Figure 3.1 in which the excitation sequence $\mathbf{x}(n)$ is modified by a filter $W(z)$ defined using a finite number of filter coefficients w_i, where the subscript i is an integer representing the order or length of the filter. The difference between the system output $y(n)$ and the desired response $d(n)$ is the error signal defined as $e(n) = d(n) - y(n)$. The concept of the Wiener filter states that a set of optimal coefficients w_i can be selected such that the error signal is minimized. The coefficients of the linear, time-invariant Wiener filter can be subjected to the adjustment of a specified algorithm to be adaptive to the changing state of $\mathbf{x}(n)$ in time. In practice, the adaptive algorithm is designed to optimize the mean-square value of $e(n)$ to

Control of Cutting Vibration and Machining Instability: A Time-Frequency Approach for Precision, Micro and Nano Machining, First Edition. C. Steve Suh and Meng-Kun Liu.
© 2013 John Wiley & Sons, Ltd. Published 2013 by John Wiley & Sons, Ltd.

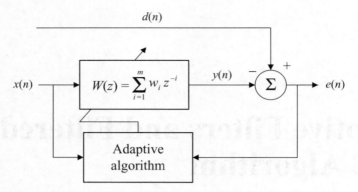

Figure 3.1 Schematic of an adaptive filter

update w_i, the filter coefficients. Modification of w_i is progressively executed every i-number of samples at a time. The adaptation process, though automatic, inevitably requires the use of a mathematically involved nonlinear regression scheme to optimize the error signal, which is a multivariable function whose dimension is determined by the order of the filter i.

3.1.1 Performance Measure

The adaptive FIR filter system in Figure 3.1 defines the filter output as a convolution sum of the filter input and the impulse response of the filter as

$$y(n) = X(z)W(z) = \sum_{k=1}^{m} w_k(n) x(n - k + 1) \tag{3.1}$$

where $X(z)$ and $W(z)$ are the z-transforms of the reference $x(n)$ and the adaptive filter, respectively. In the schematic z^{-i} is understood as representing a delay of i-number of sampling intervals. Let's introduce an Index of Performance, $J(\mathbf{w})$, as a measure for the effectiveness of minimization,

$$J(\mathbf{w}) = E[e^2(n)] = E[e(n) \cdot \bar{e}(n)] \tag{3.2}$$

where the adaptive filter coefficient vector $\mathbf{w}$ is of the order of m

$$\mathbf{w}^{\mathrm{T}} = [w_1, w_2, \ldots, w_m] \tag{3.3}$$

Thus the optimal set of w_i can be obtained in the minimum mean-square error sense by minimizing the function $J(\mathbf{w})$.

Alternatively Equation (3.1) can be expressed using the corresponding vectors

$$\begin{aligned} y(n) &= \mathbf{w}^{\mathrm{T}}\mathbf{x}(n) \\ &= \mathbf{x}^{\mathrm{T}}(n)\mathbf{w} \end{aligned} \tag{3.4}$$

Assuming that $\mathbf{w}$ is a deterministic sequence, the estimation error between the output and desired response is

$$e(n) = d(n) - \mathbf{w}^T \mathbf{x}(n) \tag{3.5}$$

The performance function is expanded using the estimation error

$$\begin{aligned}
J(\mathbf{w}) &= E[e^2(n)] \\
&= E[(d(n) - \mathbf{w}^T \mathbf{x}(n)) \cdot (d(n) - \mathbf{w}^T \mathbf{x}(n))] \\
&= E[d^2(n)] - \mathbf{w}^T E[\mathbf{x}(n)d(n)] - E[d(n)\mathbf{x}^T(n)]\mathbf{w} + \mathbf{w}^T E[\mathbf{x}(n)\mathbf{x}^T(n)]\mathbf{w}
\end{aligned} \tag{3.6}$$

Recognizing that the second and third terms are equivalent, Equation (3.6) can be arranged as follows

$$\begin{aligned}
J(\mathbf{w}) &= E[d^2(n)] - \mathbf{w}^T \mathbf{p} - \mathbf{p}^T \mathbf{w} + \mathbf{w}^T \mathbf{R} \mathbf{w} \\
&= E[d^2(n)] - 2\mathbf{p}^T \mathbf{w} + \mathbf{w}^T \mathbf{R} \mathbf{w}
\end{aligned} \tag{3.7}$$

where $\mathbf{p}$ is the cross-correlation vector associating the input $\mathbf{x}(n)$ to the desired response $d(n)$

$$\begin{aligned}
\mathbf{p} &= E[\mathbf{x}(n)d(n)] \\
&= [p(0), \; p(1), \ldots, \; p(m-1)]^T
\end{aligned} \tag{3.8}$$

with the components p being the cross-correlation function defined as

$$p(k-1) = E[\mathbf{x}(n-k+1)d(n)], \quad k = 1, \, 2, \ldots, \, m \tag{3.9}$$

and $\mathbf{R}$ is the autocorrelation matrix of the input sequence $\mathbf{x}(n)$

$$\begin{aligned}
\mathbf{R} &= E[\mathbf{x}(n)\mathbf{x}^T(n)] \\
&= \begin{bmatrix}
r_{1,1} & r_{1,2} & \cdots & r_{1,m} \\
r_{2,1} & r_{2,2} & \cdots & r_{2,m} \\
\vdots & \vdots & & \vdots \\
r_{m,1} & r_{m,2} & \cdots & r_{m,m}
\end{bmatrix}
\end{aligned} \tag{3.10}$$

where $r_{i,j} = r_{j,i}$ and

$$r_{i,j}(k-1) = E[x(n-k+1)x(n)], \quad k = 1, \, 2, \ldots, \, m \tag{3.11}$$

It is noted that the first term on the right-hand side of Equation (3.7) is the variance of the desired response $d(n)$ having a zero mean. This condition necessarily implies that the process is stationary. When $x(n)$ and $d(n)$ are jointly stationary, the index of performance $J(\mathbf{w})$ is a second-order quadratic function of the filter coefficient vector $\mathbf{w}$. The geometry of the function is commonly called the *error-performance surface*. In other words the mean-square error values described by $J(\mathbf{w})$ are defined in an m-dimensional space with a global minimum. Lastly the autocorrelation matrix $\mathbf{R}$ has to be positive-semidefinite to have a unique minimum in the space defined by $\mathbf{w}$. That is, all eigenvalues of $\mathbf{R}$ are required to be real and non-negative [2].

3.1.2   Optimization of Performance Function

When the performance function $J(\mathbf{w})$ is minimized in the mean-square sense, the corresponding $\mathbf{w}$ is optimal in keeping the error $e(n) = d(n) - y(n)$ small in the resultant FIR filtering operation. To obtain the optimal set of filter coefficients $\mathbf{w}_0$, the gradient of $J(\mathbf{w})$ is considered as follows

$$\nabla J(\mathbf{w}) = \frac{\partial J(\mathbf{w})}{\partial \mathbf{w}} = \mathbf{0} \tag{3.12}$$

where the vector of the gradient operator ∇ is

$$\nabla = \left[\begin{array}{cccc} \dfrac{\partial}{\partial w_1} & \dfrac{\partial}{\partial w_2} & \cdots & \dfrac{\partial}{\partial w_m} \end{array} \right]^{\mathrm{T}} \tag{3.13}$$

Since

$$\frac{\partial J(\mathbf{w})}{\partial \mathbf{w}} = \frac{\partial (E[d^2(n)])}{\partial \mathbf{w}} - \frac{\partial (2\mathbf{p}^{\mathrm{T}}\mathbf{w})}{\partial \mathbf{w}} + \frac{\partial (\mathbf{w}^{\mathrm{T}}\mathbf{R}\mathbf{w})}{\partial \mathbf{w}} = \mathbf{0} \tag{3.14}$$

and

$$\frac{\partial (E[d^2(n)])}{\partial \mathbf{w}} = \mathbf{0} \tag{3.15}$$

$$\frac{\partial (2\mathbf{p}^{\mathrm{T}}\mathbf{w})}{\partial \mathbf{w}} = 2\mathbf{p} \tag{3.16}$$

$$\frac{\partial (\mathbf{w}^{\mathrm{T}}\mathbf{R}\mathbf{w})}{\partial \mathbf{w}} = 2\mathbf{R}\mathbf{w} \tag{3.17}$$

as a result the optimal $\mathbf{w}_0$ that satisfies Equation (3.14) has the following property

$$-2\mathbf{p} + 2\mathbf{R}\mathbf{w}_0 = \mathbf{0} \tag{3.18}$$

or

$$\mathbf{R}\mathbf{w}_0 = \mathbf{p} \tag{3.19}$$

Known as the Wiener–Hopf equation, Equation (3.19) can be alternatively expressed as

$$\sum_{j=1}^{m} r_{i,j} w_j = p_i, \quad i = 1, 2, \ldots, m \tag{3.20}$$

Assuming that $\mathbf{R}^{-1}$ exists, the optimal vector $\mathbf{w}_0$ is determined through the availabilities of the cross-correlation vector $\mathbf{p}$ and the autocorrelation matrix $\mathbf{R}$

$$\mathbf{w}_0 = \mathbf{R}^{-1}\mathbf{p} \tag{3.21}$$

The minimum value of $J(\mathbf{w})$ is then

$$J_{\min}(\mathbf{w}) = J(\mathbf{w}_0) = E[d^2(n)] - 2\mathbf{p}^{\mathrm{T}}\mathbf{w}_0 + \mathbf{w}_0^{\mathrm{T}}\mathbf{R}\,\mathbf{w}_0 \tag{3.22}$$

or

$$J_{\min}(\mathbf{w}) = J(\mathbf{w}_0) = E[d^2(n)] - \mathbf{P}^{\mathrm{T}}\mathbf{R}^{-1}\,\mathbf{P} \tag{3.23}$$

Equations (3.22) and (3.23) provide one approach for the design of optimal Wiener filters. Other alternative designs can be formulated starting with considering

$$\frac{\partial J(\mathbf{w})}{\partial w_i} = E\left[2e(n)\frac{\partial e(n)}{\partial w_i}\right] \tag{3.24}$$

Recalling that $e(n) = d(n) - y(n)$ and that the desired response $d(n)$ is independent of the filter vector $\mathbf{w}$, one has

$$\begin{aligned}
\frac{\partial e(n)}{\partial w_i} &= \frac{\partial d(n)}{\partial w_i} - \frac{\partial y(n)}{\partial w_i} \\
&= -\frac{\partial y(n)}{\partial w_i} \\
&= -x(n-i)
\end{aligned} \tag{3.25}$$

Equation (3.24) becomes

$$\frac{\partial J(\mathbf{w})}{\partial w_i} = -2E[e(n)x(n-i)], \quad i = 1, 2, \ldots, m \tag{3.26}$$

Denoting $e_0(n)$ as the estimation error corresponding to the case when the performance function is minimized and thus $\mathbf{w}_0$ is found,

$$\left.\frac{\partial J(\mathbf{w})}{\partial w_i}\right|_{\mathbf{w}=\mathbf{w}_0} = -2E[e_0(n)x(n-i)] = 0, \quad i = 1, 2, \ldots, m \tag{3.27}$$

Using Equation (3.1) it can be shown that

$$E[e_0(n)y_0(n)] = 0, \quad i = 1, 2, \ldots, m \tag{3.28}$$

where y_0 is the filtered optimal output corresponding to $\frac{\partial J(\mathbf{w})}{\partial w_i} = 0$. The proper interpretation of the two equations above is that when the Wiener filter coefficients are set to their optimal values in the mean-square error sense, the input sequence and the optimized Wiener filter output are both uncorrelated with the estimation error. $x(n - i)$ and $y_0(n)$ are said to be orthogonal to $e_0(n)$. This property of orthogonality has been favorably explored to realize better computational efficiency for updating the adaptive filter when the inputs are nonstationary.

3.2 Gradient Descent Optimization

Updating the filter with the optimal set of $\mathbf{w}_0$ at each filtering and error estimation operation requires the Wiener–Hopf equation in Equation (3.19) to be solved. However, the task can be daunting as well as computationally costly when the input data rate is high and the filter length (number of filter coefficients) is large. Mathematical optimization such as the Gradient Descent method is often the tool of choice for finding the global minimum of the error performance surface which is the optimal solution to the Wiener adaptive filter problem [3]. Also named as steepest descent, gradient descent is a first-order optimization method. The solution strategy of the method is to search for and follow the paths whose direction gradients are the steepest in the negative sense. This is progressively performed following an iterative scheme till the global minimum is reached, at which point the solution vector is of zero gradient [4].

To illustrate the iterative procedure, a function $f(\mathbf{x})$ is considered. The goal is to find a local extremum in the neighborhood of an arbitrary point $\mathbf{x}_0$. The search is implemented by always following the direction along which $f(\mathbf{x})$ decreases most quickly, which is opposite to the direction defined by $\nabla f(\mathbf{x})$

$$\mathbf{x}_{i+1} = \mathbf{x}_i - \mu_i \nabla f(\mathbf{x}_i) \tag{3.29}$$

where μ_i is a real-valued constant that determines the step size to be taken to move from the current location at $\mathbf{x}_i$ to the next location at $\mathbf{x}_{i+1}$. The choice of μ_i is not arbitrary. Because the descending path has the minimum functional value at $\mathbf{x}_{i+1}$ where the directional gradient is zero, it is necessary that

$$
\begin{aligned}
\frac{df(\mathbf{x}_{i+1})}{d\mu_i} &= 0 \\
&= [\nabla f(\mathbf{x}_{i+1})]^{\mathrm{T}} \left(\frac{d\mathbf{x}_{i+1}}{d\mu_i} \right) \\
&= [\nabla f(\mathbf{x}_{i+1})]^{\mathrm{T}} [\nabla f(\mathbf{x}_i)]
\end{aligned}
\tag{3.30}
$$

Equation (3.30) dictates that μ_i should be so chosen that $\nabla f(\mathbf{x}_{i+1})$ and $\nabla f(\mathbf{x}_i)$ are mutually orthogonal to ensure proper solution stability and convergence.

The steepest gradient procedures outlined above can be followed to search the performance function $J(\mathbf{w})$ for the optimal filter vector solution at $\mathbf{w}_0$. The iterative solution scheme at n-th iteration is similar to Equation (3.29),

$$\mathbf{w}(n+1) = \mathbf{w}(n) - \frac{\mu}{2} \frac{\partial J(\mathbf{w})}{\partial \mathbf{w}}\bigg|_{\mathbf{w}=\mathbf{w}(n)} \tag{3.31}$$

From Equations (3.14)–(3.18)

$$\frac{\partial J(\mathbf{w})}{\partial \mathbf{w}}\bigg|_{\mathbf{w}=\mathbf{w}(n)} = -2\mathbf{p} + 2\mathbf{R}\mathbf{w}(n) \tag{3.32}$$

The solution procedure making use of the steepest-descent concept results in a recursive scheme

$$\mathbf{w}(n+1) = \mathbf{w}(n) - \mu[-\mathbf{p} + \mathbf{R}\mathbf{w}(n)] \tag{3.33}$$

where μ, the step size parameter controlling the incremental estimation of the filter vector at each iteration step, is to be selected following the criteria stated in Equation (3.30).

It is explicit from Equation (3.33) that, in addition to μ, $\mathbf{R}$, the autocorrelation matrix of the input vector $\mathbf{x}(n)$, also plays a role in dictating the stability of the iterative solution algorithm. To see how μ and $\mathbf{R}$ impact stability performance, consider the error vector

$$\mathbf{e}_\mathbf{w}(n) = \mathbf{w}(n) - \mathbf{w}_0 \tag{3.34}$$

Applying the vector to Equation (3.33) at two consecutive iterative steps, we have

$$\mathbf{e}_\mathbf{w}(n+1) = (\mathbf{I} - \mu\mathbf{R})\mathbf{e}_\mathbf{w}(n) \tag{3.35}$$

Since $\mathbf{R}$ is positive-semidefinite, real, and symmetric, the following similarity transformation on $\mathbf{R}$ exists

$$\mathbf{R} = \mathbf{Q}\Lambda\mathbf{Q}^\mathrm{T} \tag{3.36}$$

where Λ is a diagonal matrix with diagonal entries λ_i equal to the eigenvalues of $\mathbf{R}$ and the unitary matrix $\mathbf{Q}^\mathrm{T} = \mathbf{Q}^{-1}$ whose rows carry the eigenvectors corresponding to the eigenvalues of $\mathbf{R}$. Equation (3.35) now becomes

$$\mathbf{Q}^\mathrm{T}\mathbf{e}_\mathbf{w}(n+1) = (\mathbf{I} - \mu\Lambda)\mathbf{Q}^\mathrm{T}\mathbf{e}_\mathbf{w}(n) \tag{3.37}$$

Note that $\mathbf{Q}^\mathrm{T}\mathbf{e}_\mathbf{w}(n)$ is a transformation of $\mathbf{e}_\mathbf{w}$ via $\mathbf{Q}^\mathrm{T}$ having the following properties

$$\mathbf{s}(n) = \mathbf{Q}^\mathrm{T}\mathbf{e}_\mathbf{w}(n) = \mathbf{Q}^\mathrm{T}[\mathbf{w}(n) - \mathbf{w}_0] \tag{3.38}$$

$$\mathbf{s}(n+1) = (\mathbf{I} - \mu\Lambda)\mathbf{s}(n) \tag{3.39}$$

Recall that all eigenvalues of the matrix $\mathbf{R}$ are positive and real. This particular property ensures that $\mathbf{s}(n)$, the transformation of the error vector $\mathbf{e_w}(n)$, as a response in Equation (3.39), displays no numerical oscillations. Denoting λ_k as the k-th eigenvalue of the matrix $\mathbf{R}$, Equation (3.39) is alternatively expressed using its vector components

$$s_k(n+1) = (1 - \mu\lambda_k)s_k(n) \tag{3.40}$$

where $k = 1, 2, \ldots, m$ with m being the filter length. By specifying an initial $s_k(0)$, the solution obtained at the n-th iteration is simply

$$s_k(n) = (1 - \mu\lambda_k)^n s_k(0) \tag{3.41}$$

The convergence of Equation (3.41) is a direct measure of the stability of the steepest descent algorithm. The condition for Equation (3.41) to converge is determined by

$$-1 < (1 - \mu\lambda_k) < 1 \tag{3.42}$$

$s_k(n)$, the transformation of estimation error, is seen to approach zero as the iteration approaches infinity, thus rendering the approaching of $\mathbf{w}(n)$ to $\mathbf{w}_0$ the optimal solution. The condition in Equation (3.42) requires that the following constraint be satisfied by μ to establish stability:

$$0 < \mu < \frac{2}{\lambda_{\max}} \tag{3.43}$$

with $\lambda_{\max}$ being the largest eigenvalue of the autocorrelation matrix $\mathbf{R}$.

3.3   Least-Mean-Square Algorithm

The method of gradient descent introduced in the previous section assumes a well-defined performance function to iterate through to search for the optimal solutions. When such a function is not available, which is more often than not the case in the real-world, measures of the directional gradients have to be estimated from the finite number of data that are accessible at each iteration step. The Least-Mean-Square (LMS) method, along with its many variations, is widely adopted for the task of adapting to the available input data in the mean-square sense and estimating and updating the filter vector following the steepest gradient algorithm [5].

The classical LMS algorithm is a stochastic implementation of the steepest gradient method by considering the *instantaneous coarse estimate* of the performance function. This is implemented by estimating the gradient of the performance function $\nabla J(\mathbf{w})|_{\mathbf{w}(n)} = -2\mathbf{p} + 2\mathbf{R}\mathbf{w}(n)$ using the instantaneous coarse estimates of the autocorrelation matrix $\mathbf{R}$ and cross-correlation vector $\mathbf{p}$ defined as follows

$$\hat{\mathbf{R}}(n) = \mathbf{x}(n)\mathbf{x}^{\mathrm{T}}(n) \tag{3.44}$$

$$\hat{\mathbf{p}}(n) = d(n)\mathbf{x}(n) \tag{3.45}$$

where the hats denote the corresponding instantaneous coarse estimates. The instantaneous coarse estimate of the gradient of the performance function is defined using the two instantaneous estimates in Equations (3.44) and (3.45)

$$\hat{\nabla} J(\mathbf{w})|_{\mathbf{w}(n)} = -2d(n)\mathbf{x}(n) + 2\mathbf{x}(n)\mathbf{x}^{T}(n)\mathbf{w}(n) \tag{3.46}$$

By substituting Equation (3.46) into the steepest gradient recursion in Equation (3.33), a new recursive scheme is obtained as a result

$$\hat{\mathbf{w}}(n + 1) = \hat{\mathbf{w}}(n) + \mu\mathbf{x}(n)[d(n) - \mathbf{x}^{T}(n)\hat{\mathbf{w}}(n)] \tag{3.47}$$

In practice the LMS recursion in Equation (3.47) is implemented by executing the following equation pairs sequentially to update the instantaneous estimate $\hat{\mathbf{w}}(n + 1)$

$$\hat{e}(n) = d(n) - \mathbf{x}^{T}(n)\hat{\mathbf{w}}(n) \tag{3.48}$$

$$\hat{\mathbf{w}}(n + 1) = \hat{\mathbf{w}}(n) + \mu\mathbf{x}(n)\hat{e}(n) \tag{3.49}$$

in which the estimation error $\hat{e}(n)$ is determined by the current estimate of $\hat{\mathbf{w}}(n)$. When the estimation error is significant, by not meeting the set tolerance, $\mu\mathbf{x}(n)\hat{e}(n)$ is then determined using $\hat{e}(n)$ to correct the current estimate of $\hat{\mathbf{w}}(n)$ to estimate and update $\hat{\mathbf{w}}(n + 1)$ according to Equation (3.49). As evaluations of the autocorrelation matrix are not required, the LMS algorithm given in Equations (3.47)–(3.49) performs no matrix inversion.

Haykin [2] shows that the convergence of the mean-square error alone is insufficient to warrant stability of the LMS algorithm if the variance of the filter coefficients $E[d^{2}(n)]$ in Equation (3.7) is unbounded. This can be illustrated by considering the following:

$$\hat{\mathbf{w}}(n + 1) - \hat{\mathbf{w}}(n) = \mu\mathbf{x}(n)\hat{e}(n) \tag{3.50}$$

Even if the adaptive step μ is properly chosen and the estimate $\hat{\mathbf{w}}(n)$ is close to the optimal coefficient vector $\mathbf{w}_0$ so that $\hat{\mathbf{w}}(n) \approx \mathbf{w}_0$, because $\mathbf{x}(n)\hat{e}(n)$ is random, the updating of the weights of the filter can never ideally achieve zero. That is,

$$\hat{\mathbf{w}}(n + 1) - \mathbf{w}_0 = \mu\mathbf{x}(n)\hat{e}(n) \neq \mathbf{0} \tag{3.51}$$

It is evident from Equation (3.51) that $\hat{\mathbf{w}}(n + 1)$ never will really settle into $\mathbf{w}_0$. In fact it will always oscillate about $\mathbf{w}_0$.

Assume that the reference sequence $d(n)$ carries a zero-mean independent noise sequence $\varepsilon(n)$ as follows

$$d(n) = [\mathbf{w}_0]^{T}\mathbf{x}(n) + \varepsilon(n) \tag{3.52}$$

so that when $\hat{\mathbf{w}}(n) = \mathbf{w}_0$, the estimation error defined in Equation (3.5) becomes

$$e(n) = [\mathbf{w}_0]^{T}\mathbf{x}(n) + \varepsilon(n) - \mathbf{w}^{T}\mathbf{x}(n) \tag{3.53}$$

As such, the corresponding mean-squares are

$$E[e^2(n)] = E[\varepsilon^2(n)] = J_{min}(\mathbf{w}) \tag{3.54}$$

Since, in practice, $\hat{\mathbf{w}}(n) \neq \mathbf{w}_0$, it can be concluded from Equation (3.54) that

$$E[e^2(n)] > J_{min}(\mathbf{w}) \tag{3.55}$$

As a measure for how much the weight vector is misadjusted from its optimal setting, the *misadjustment factor* is defined as follows

$$M = \frac{E[e^2(n)] - J_{min}(\mathbf{w})}{J_{min}(\mathbf{w})} \tag{3.56}$$

An equivalent expression for the adjustment factor derived based on the noise covariance matrix is available [6]:

$$M = \frac{\mu \, \text{trace}(\mathbf{R})}{1 - \mu \, \text{trace}(\mathbf{R})} \tag{3.57}$$

Using Equations (3.56) and (3.57) the variance can be controlled to be finite and bounded. This is done by choosing an adaptive step μ and using it to select a desired M to realize a specified (designed) deviation from the optimal Wiener filter.

3.4   Filtered-x LMS Algorithm

Adaptive filters running on the LMS algorithm, such as the one depicted in Figure 3.1, are widely employed as controllers in many engineering applications. For example, adaptive filters are essential components in active noise control for the cancellation of acoustic disturbance or the compensation of sound quality, or both. Modern day smart structures and bridges also exploit LMS-based active controllers to offset detrimental seismic trembles. Adaptive filters are usually incorporated in real-world configurations conceptually similar to the feedforward scheme depicted in Figure 3.2, in which the filter tracks system variation, estimates input, adapts to response by updating the filter weights, and exerts proper compensation to facilitate control of the plant.

The configuration in the schematic is one of system identification, where the LMS-based adaptive filter $W(z)$ estimates the response of the unknown plant. The plant is excited by an input $x(n)$ and characterized using a transfer function $P(z)$. Response estimation is established through $F(z)$. Since it is an integral part of the plant, whose physical implementation could be an actuator or a sensory array or a nonautonomous subsystem, it is necessary that $F(z)$ is compensated for. Upon successful adaptation to the dynamics of the unknown plant through updating the filter coefficients and minimizing the residual error $e(n)$ in the mean-square sense, control is applied. The path defined by $X(z)P(z)$ is commonly called the primary; while the secondary path is by $X(z)W(z)F(z)$.

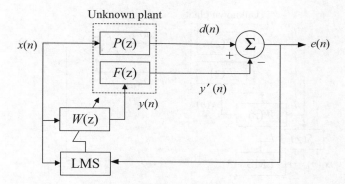

Figure 3.2   LMS-based adaptive system identification configuration

The frequency domain representation of the error $e(n)$ is simply

$$
\begin{aligned}
E(z) &= D(z) - Y'(z) \\
&= X(z)P(z) - X(z)W(z)F(z) \\
&= [P(z) - W(z)F(z)]X(z)
\end{aligned}
\tag{3.58}
$$

Assume that the adaptive filter is of sufficient order to ensure convergence and reach zero residual error as a result; that is, when the coefficient vector $\mathbf{w}(n) = \mathbf{w}_0$ and $E(z) = 0$. When this is the case, $W(z)$ is required to be an optimal transfer function that models the primary path $P(z)$ and inversely models the secondary path $F(z)$ *simultaneously* as follows

$$
[W(z)]_{\mathbf{w}=\mathbf{w}_0} = \frac{P(z)}{F(z)}
\tag{3.59}
$$

Equation (3.59) presents three major concerns over the feasibility of such a configuration in Figure 3.2 [5]. All of them are closely related to the underlying attributes of the secondary path $F(z)$. The first is that it is impossible to invert the inherent delay caused by $F(z)$ if the primary path $P(z)$ does not contain a delay at least of the same time scale. The second is the system would become mathematically singular and physically unstable if $F(\omega_c) = 0$ at an uncontrollable frequency ω_c. Last is the scenario when $P(\omega_o) = 0$, in which the unobservable control frequency ω_o causes a zero in the primary path that "nulls" the filter.

Many propositions for compensating the undesirable effects of the secondary-path transfer function $F(z)$ have been made since the early 1980s. Among them, the approach that addressed the problems by modifying the adaptive LMS algorithm to ensure convergence has been most popular [6, 7, 8, 9]. Secondary path effects are mitigated by running the input sequence $x(n)$ through an auxiliary estimate filter $S(z)$ which is placed along the secondary path as shown in Figure 3.3. As the input is being filtered in the adaptive architecture, the modified algorithm is thus accordingly named a Filtered-x LMS algorithm (FXLMS).

The dynamic $y'(n)$-to-$d(n)$ correlation in Figure 3.3 is estimated by $F(z)$ whose filter coefficients are obtained and updated by the offline LMS algorithm. The output of the control

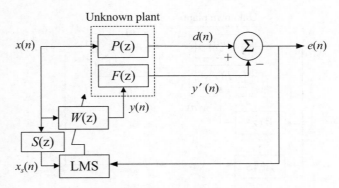

Figure 3.3   Filtered-x LMS structure

filter $y(n)$, which is also the input to the control system, is a convolution sum of the impulse response of the filter $F(z)$ and the external input $x(n)$

$$y(n) = \sum_{i=0}^{m-1} w_i(n)x(n-i) \qquad (3.60)$$

where m is the order of the adaptive filter $W(z)$. The output $y'(n)$ from the k-th order $F(z)$ due to the control input $y(n)$ is

$$y'(n) = \sum_{j=0}^{k-1} f_j y(n-j) \qquad (3.61)$$

Substituting Equation (3.60) into Equation (3.61) we have

$$y'(n) = \sum_{j=0}^{k-1} f_j \sum_{i=0}^{m-1} w_i(n)x(n-i-j) \qquad (3.62)$$

At the sum junction, the output error $e(n)$ is

$$e(n) = d(n) - y'(n)$$

$$= d(n) - \sum_{j=0}^{k-1} f_j \sum_{i=0}^{m-1} w_i(n)x(n-i-j) \qquad (3.63)$$

Assuming that the adaptive FIR filter and the forward path $F(z)$ commute, then the order of summation in Equation (3.63) can be rearranged to allow the net output of the system to be approximated in an alternative form as follows

$$e(n) = d(n) - \sum_{i=0}^{m-1} w_i(n) \sum_{j=0}^{k-1} f_j x(n-i-j)$$

$$\approx d(n) - \sum_{i=0}^{m-1} w_i(n) x'(n-i)$$

(3.64)

where

$$x'(n-i) = \sum_{j=0}^{k-1} f_j x(n-i-j)$$

(3.65)

The validity of Equation (3.64) requires that the adaptive filter varies on a time scale that is slower than that of the impulse response of the forward path. The approximation error would be non-negligible and significant if the principal assumption is violated. However, it has been reported that, in practical applications, the FXLMS algorithm could demonstrate stability when the adaptive coefficients vary within the same time scale associated with the dynamic response of the forward path [7, 10].

The gradient of the performance function in Equation (3.32) is

$$\left[\frac{\partial J(\mathbf{w})}{\partial \mathbf{w}}\right]_{\mathbf{w}=\mathbf{w}(n)} = \frac{\partial J(n)}{\partial w_i(n)}$$

$$= -2x'(n-i)e(n)$$

(3.66)

The coefficients of the adaptive filter are updated according to the recursive scheme defined in Equation (3.31)

$$w_i(n+1) = w_i(n) + \mu x'(n-i)e(n)$$

(3.67)

Thus $y'(n)$ is adapted at j-th sampling step using the $x'(n-i)$ calculated in Equation (3.65) and the coefficient vector $\mathbf{w}$ updated by Equation (3.67) to minimize the output error. Equations (3.65) and (3.67) indicate that the input sequence $x(n)$ is filtered by $F(z)$ before updating the filter coefficients w_i. This is realized in practice by introducing an additional filter $S(z)$ to estimate $F(z)$, thus

$$x_s(n) = \sum_{i=0}^{m-1} s_i(n) x(n-i)$$

(3.68)

and the updating of $\mathbf{w}$ in Equation (3.67) becomes

$$w_i(n+1) = w_i(n) + \mu x_s(n-i)e(n)$$

(3.69)

or alternatively

$$\mathbf{w}(n + 1) = \mathbf{w}(n) + \mu e(n)\mathbf{S}^{\mathrm{T}}\mathbf{x} \tag{3.70}$$

where the filtering matrix $\mathbf{S}$ is defined as

$$\mathbf{S} = \begin{bmatrix} s_0 & 0 & \cdots & & \cdots & 0 \\ s_1 & s_0 & 0 & \cdots & & 0 \\ \vdots & s_1 & \ddots & \ddots & & \vdots \\ s_k & \vdots & & \ddots & \ddots & 0 \\ 0 & s_k & & & \ddots & s_0 \\ \vdots & & & \ddots & & s_1 \\ \vdots & & & & \ddots & \vdots \\ 0 & \cdots & & \cdots & 0 & s_k \end{bmatrix} \tag{3.71}$$

where s_i is understood to be the impulse response of the auxiliary filter $S(z)$ of the order of k.

Note that one of the possible configurations meeting the condition for stability and convergence is when the estimate of the secondary path equals the primary path $F(z)$; that is, $S(z) = F(z)$. However, such a case is rare in practical applications because the control transfer function $F(z)$ is usually unknown, thus the need for the adaptive scheme to adapt to the dynamics of the plant.

Unlike the classic LMS algorithm, the maximum step size μ_{max} associated with the FXLMS algorithm is a simultaneous function of the adaptive filter length, filtered reference signal, and delays in the forward path $F(z)$ [11]. Instability may also come from model errors that could render the FXLMS update rule highly unstable [12]. It is common to address the particular problem by improving the model of the system using elaborated system identification methods. However, model uncertainties due to prominent system nonlinearity would still degrade the performance and destabilize the update rule. As have discussed in Section 3.3, the improper choice of the step size μ could have a negative impact on the misadjustment factor M. A large μ usually gives a better performance in tracking nonstationary signals. But in doing so it also generates large misadjustment noise, thus risking inducing instability in the adaptive filter.

References

[1] Wiener, N., 1949, *Extrapolation, Interpolation, and Smoothing of Stationary Time Series*, John Wiley and Sons, New York.
[2] Haykin, S., 1996, *Adaptive Filter Theory*, 3rd edn, Englewood Cliffs, New Jersey, Prentice Hall.
[3] Elliot, S. J., 2001, *Signal Processing for Active Control*, London, Academic Press.
[4] Bonnans, J. D., 2003, *Numerical Optimization: Theoretical and Practical Aspects*, 2nd edn, Spring-Verlag, Berlin Heidelberg.
[5] Kuo, S. M., Morgan, D. R., 1996, *Active Noise Control Systems: Algorithms and DSP Implementations*, John Wiley and Sons, New York.
[6] Widrow, B., Walach, E., 2007, *Adaptive Inverse Control: A Signal Processing Approach*, Reissue edn, John Wiley and Sons, New York.

[7] Widrow, B., Stearns, S. D., 1985, *Adaptive Signal Processing*, Englewood Cliffs, New Jersey, Prentice Hall.

[8] Morgan, D. R., 1980, "An Analysis of Multiple Correlation Cancellation Loops With A Filter in The Auxiliary Path," *IEEE Transactions on Acoustics, Speech, and Signal Processing*, ASSP-28, Aug. 1980, 454–67.

[9] Burgess, J. C., 1981, "Active Adaptive Sound Control in A Duct: A Computer Simulation," *Journal of American Acoustic Society*, 70(3), 715–26.

[10] Nelson, P. A., Elliott, S. J., 1992, *Active Control of Sound*, London, Academic Press.

[11] Elliott, S. J., Nelson, P.A., 1989, "Multiple-Point Equalization in A Room Using Adaptive Digital Filters," *Journal of Audio Engineering Society*, 37(11), 899–907.

[12] Fraanje, R., Verhaegen, M., Elliott, S. J., 2007, "Robustness of The Filtered-X LMS Algorithm—Part I: Necessary Conditions for Convergence and The Asymptotic Pseudospectrum of Toeplitz Matrices," *IEEE Transactions on Signal Processing*, 55(8), 4029–37.

4

Time-Frequency Analysis

The need for representing a nonlinear, nonstationary signal in the joint time-frequency domain was indicated in Section 2.5 in which the inadequacy of popular characterization tools were briefly discussed. The Gabor wavelet, which provides the optimal simultaneous time-frequency resolution available of all wavelet functions, was shown to be unsatisfactory in resolving a milling vibration signal for indications of perturbed cutting state and deteriorating milling stability. One of the conclusions made in that section was that all Fourier-based methodologies, including the wavelet transform, are not viable for differentiating the evolution of spectral components in time indicative of bifurcated state of motion or imminent dynamic instability.

The fundamental properties of the sinusoidal basis function of the Fourier transform are the primary reason for the poor resolution. To see why, consider the Fourier transform of the time function $f(t) \in L^2(\mathbf{R})$ in Equation (2.13). Because the exponential basis function $e^{-i\omega t} = \sin(\omega t) - i\cos(\omega t)$ is defined on $(-\infty, \infty)$ in the t-domain, and also mapped into $(-\infty, \infty)$ in the ω-domain, the infinite integral requires that $f(t)$ be fully defined and available on the infinite t-domain to have a proper valid representation of the function in the ω-domain. That is, a perturbation experienced by $f(t)$ in the t-domain is interpreted in the entire ω-domain as a global effect. This loss of t-domain locality as a result in the transformed domain is further illustrated using the Dirac delta function $\delta(t)$ as follows:

$$\int_{-\infty}^{\infty} \delta(t-y)f(t)dt = f(y) \tag{4.1}$$

The Fourier transform of $\delta(t)$ is

$$\hat{\delta}(\omega) = \int_{-\infty}^{\infty} \delta(t)e^{-i\omega t}dt = 1 \tag{4.2}$$

Thus an impulse in the t-domain is not represented as a corresponding impulse in the ω-domain. It is delineated as an infinite Fourier spectrum. As anomaly and perturbation in a nonstationary signal are often physically short impulses of negligible amplitude, their effects on the overall dynamics of the signal are inevitably obscured by the presence of infinite numbers of ω components.

In addition to misrepresentation in the Fourier domain, a perturbed dynamic response can also be misinterpreted. Observe again Equation (2.13) in which the transformation is an operation of mathematically averaging the modulated function $f(t)e^{-i\omega t}$ in the infinite integral sense. To satisfy the mandatory and also rigid requirement for generating an average, fictitious ω components are unavoidable. Misrepresentations by misleading frequency components are most prominent in nonlinear, nonstationary signals [1] when Fourier transform is applied.

One way to establish the local frequency content corresponding to a local time event is to allow time to be corroborated locally in the transform domain. Reference to time can be established by introducing a well-defined "window function" into the Fourier integral. Such an idea was inspired by the Wiener–Khinchin Theorem [2] which states that

$$|F(\omega)|^2 = \int_{-\infty}^{\infty} R(\tau)e^{-i\omega\tau}d\tau \tag{4.3}$$

where $|F(\omega)|^2$ is the power spectrum of the function $f(t)$ and $R(\tau)$ is the autocorrelation function of $f(t)$ defined as

$$R(\tau) = \int_{-\infty}^{\infty} f(t)\bar{f}(t-\tau)dt \tag{4.4}$$

where $\bar{f}(t)$ is the complex conjugate of $f(t)$. Thus, the power spectrum of $f(t)$ is alternatively expressed using its autocorrelation function. By subjectively making the autocorrelation function t-dependent as $R(\tau, t)$, an additional t-dimension is imparted into Equation (4.3) as follows

$$|F(\omega, t)|^2 = \int_{-\infty}^{\infty} R(\tau, t)e^{-i\omega\tau}d\tau \tag{4.5}$$

Evaluating ω with a reference to t established by the window function $R(\tau, t)$, Equation (4.5) provides the basic framework for all the integral-based time-frequency representation methods, including the short-time Fourier transform (STFT) and continuous wavelet transform (CWT).

4.1 Time and Frequency Correspondence

As is implied by its name, time-frequency control addresses the control of vibrations in the simultaneous time-frequency domain. A comprehensive understanding of the correspondence between time and frequency defined in the context of the Fourier transform would prove beneficial for establishing the notion of time-frequency control. We will begin by considering

$f(t)$ and $g(t)$ in the following along with their corresponding Fourier transforms $F(\omega)$ and $G(\omega)$, respectively.

If for some scalars a and b there exists a function which is a linear sum of $f(t)$ and $g(t)$ such that $s(t) = a\, f(t) \pm b\, g(t)$, then we have the following property of *linearity*,

$$
\begin{aligned}
S(\omega) &= \int_{-\infty}^{\infty} s(t)\, e^{-i\omega t}\, dt \\
&= a \int_{-\infty}^{\infty} f(t)\, e^{-i\omega t}\, dt + b \int_{-\infty}^{\infty} g(t)\, e^{-i\omega t}\, dt \\
&= a\, F(\omega) + b\, G(\omega)
\end{aligned}
\tag{4.6}
$$

A shift in the t-domain by t_0 corresponds to a multiplication by a phase factor in the ω-domain:

$$
\begin{aligned}
F_0(\omega) &= \int_{-\infty}^{\infty} f(t - t_0)\, e^{-i\omega t}\, dt \\
&= \int_{-\infty}^{\infty} f(\tau)\, e^{-i\omega(\tau + t_0)}\, d\tau \\
&= e^{-i\omega t_0} F(\omega)
\end{aligned}
\tag{4.7}
$$

Conversely a shift in the ω-domain by ω_0 corresponds to a modulation by an exponential function in the t-domain:

$$
\begin{aligned}
f_0(t) &= \frac{1}{2\pi} \int_{-\infty}^{\infty} F(\omega - \omega_0)\, e^{i\omega t}\, d\omega \\
&= \frac{1}{2\pi} \int_{-\infty}^{\infty} F(\tau)\, e^{i(\tau + \omega_0)t}\, d\tau \\
&= e^{i\omega_0 t} f(t)
\end{aligned}
\tag{4.8}
$$

Scaling in the t-domain results in inverse scaling in the ω-domain:

$$
\begin{aligned}
F_k(\omega) &= \int_{-\infty}^{\infty} f(kt)\, e^{-i\omega t}\, dt \\
&= \int_{-\infty}^{\infty} f(\tau)\, e^{-i\omega\left(\tau/k\right)}\, d\left(\frac{\tau}{k}\right) \\
&= \frac{1}{|k|} F\left(\frac{\omega}{k}\right)
\end{aligned}
\tag{4.9}
$$

When the scaling factor $k > 1$, the center frequency decreases and the corresponding spectrum is contracted, all by $1/k$ amount. When $k < 1$, the opposite is true. It can also be shown that for a nonzero constant m,

$$\begin{aligned} f_m(t) &= \frac{1}{2\pi} \int_{-\infty}^{\infty} F(m\omega)\, e^{i\omega t}\, d\omega \\ &= \frac{1}{|m|}\, f\left(\frac{t}{m}\right) \end{aligned} \tag{4.10}$$

These scaling properties are explored by employing wavelet functions to achieve controlled ω-domain or t-domain resolution in the joint time-frequency domain.

Consider the n-th derivative of $f(t)$, the corresponding Fourier transform is straightforward

$$\begin{aligned} F_n(\omega) &= \int_{-\infty}^{\infty} \left[\frac{d^n}{d\,t^n} f(t)\right] e^{-i\omega t}\, dt \\ &= \frac{d^n}{d\,t^n}\left(\int_{-\infty}^{\infty} f(t)e^{-i\omega t}\, dt\right) \\ &= (i\,\omega)^n\, F(\omega) \end{aligned} \tag{4.11}$$

Note that because the infinite integral is a linear transformation, the order of differentiation and integration is interchangeable. Equation (4.11) will be referenced later in the section on uncertainty principle.

Recall from Section 2.4 that the convolution of $f(t)$ and $g(t)$ is

$$f(t) * g(t) = \int_{-\infty}^{\infty} f(\tau)g(t-\tau)d\tau \tag{4.12}$$

and the Fourier transform of Equation (4.12) is a product in the ω-domain

$$\begin{aligned} \mathbf{F}[f(t) * g(t)] &= \int_{-\infty}^{\infty} \left\{\int_{-\infty}^{\infty} f(\tau)g(t-\tau)d\tau\right\} e^{-i\omega t}\, dt \\ &= F(\omega) \cdot G(\omega) \end{aligned} \tag{4.13}$$

When $t = 0$ we have

$$\int_{-\infty}^{\infty} f(\tau)g(-\tau)d\tau = \frac{1}{2\pi} \int_{-\infty}^{\infty} F(\omega)G(\omega)d\omega \tag{4.14}$$

In the case that $f(t)$ is complex and $f(t) = g(-t)$ we have *Parseval's Theorem* as a result

$$\int_{-\infty}^{\infty} |f(t)|^2 dt = \frac{1}{2\pi} \int_{-\infty}^{\infty} |F(\omega)|^2 d\omega \tag{4.15}$$

or alternatively

$$\langle f(t), \ \bar{f}(t) \rangle = \frac{1}{2\pi} \langle F(\omega), \ \bar{F}(\omega) \rangle = E \tag{4.16}$$

Equations (4.15) and (4.16) imply that the total energy is conserved and the energies of the signal determined in the time and frequency domain are equivalent and interchangeable. This particular conclusion has significant implications for the development of the concept of nonlinear time-frequency control in Chapter 7.

4.2 Time and Frequency Resolution

With the basic notions of signal energy and energy conservation established, the temporal and spectral resolutions of a function can now be defined using the concepts of standard deviation and root-mean-square. Apply the conservation of signal energy in Equations (4.15) and (4.16) to determine the *mean time* $\langle t \rangle$ and *mean frequency* $\langle \omega \rangle$ below,

$$\langle t \rangle = \frac{1}{E} \int_{-\infty}^{\infty} t \, |f(t)|^2 dt \tag{4.17}$$

and

$$\langle \omega \rangle = \frac{1}{2\pi E} \int_{-\infty}^{\infty} \omega \, |F(\omega)|^2 d\omega \tag{4.18}$$

If $f(t)$ is finite and of compact support; that is, $f(t) = 0$ outside a finite set on $(-\infty, \infty)$, $f(t)$ is often called a time-frequency window function having $\langle t \rangle$ and $\langle \omega \rangle$ as its centers in the time and the transformed frequency domains, respectively. The corresponding time duration and frequency bandwidth are defined in the root-mean-square sense using the two centers as

$$(\Delta_t)^2 = \frac{1}{E} \left[\int_{-\infty}^{\infty} (t - \langle t \rangle)^2 \, |f(t)|^2 dt \right] = \int_{-\infty}^{\infty} \frac{t^2}{E} \, |f(t)|^2 dt - \langle t \rangle^2 \tag{4.19}$$

and

$$(\Delta_\omega)^2 = \frac{1}{2\pi E} \left[\int_{-\infty}^{\infty} (\omega - \langle \omega \rangle)^2 \, |F(\omega)|^2 d\omega \right] = \frac{1}{2\pi} \int_{-\infty}^{\infty} \frac{\omega^2}{E} \, |F(\omega)|^2 d\omega - \langle \omega \rangle^2 \tag{4.20}$$

Equations (4.19) and (4.20) are the two variances quantitatively describing the spreads of the signal with respect to $\langle t \rangle$ and $\langle \omega \rangle$.

4.3 Uncertainty Principle

There is a very important property associated with the time-frequency width product of the window function $\Delta_t \Delta_\omega$, which is bounded by the uncertainty principle that states that if $|t| \to \infty$, $\{f(t)\sqrt{t}\} \to 0$ then

$$\Delta_t \Delta_\omega \geq \frac{1}{2} \qquad (4.21)$$

The equality holds only when $f(t)$ is of the Gaussian type, such as $f(t) = Ae^{-\alpha t^2}$.

Equation (4.21) implies that time and frequency localizations cannot be simultaneously achieved with arbitrary resolution. That is, using a window function to attempt a fine temporal resolution with a small time duration will inherently result in a coarse spectral resolution with a large frequency bandwidth. Thus the selection of Δ_t is dependent upon Δ_ω, and vice versa.

To show the uncertainty principle with the inequality, we first assume that $\langle t \rangle = 0$ and $\langle \omega \rangle = 0$ for simplicity. The corresponding widths defined in Equations (4.19) and (4.20) are then

$$(\Delta_t)^2 = \int_{-\infty}^{\infty} t^2 |f(t)|^2 dt \qquad (4.22)$$

and

$$(\Delta_\omega)^2 = \frac{1}{2\pi} \left[\int_{-\infty}^{\infty} \omega^2 |F(\omega)|^2 d\omega \right] \qquad (4.23)$$

so that

$$(\Delta_t \Delta_\omega)^2 = \left(\left[\int_{-\infty}^{\infty} t^2 |f(t)|^2 dt \right] \right) \left(\frac{1}{2\pi} \left[\int_{-\infty}^{\infty} \omega^2 |F(\omega)|^2 d\omega \right] \right) \qquad (4.24)$$

Assume a function $g(t)$ whose Fourier transform is $G(\omega) = \omega F(\omega)$. By applying Parseval's Theorem and using the two time domain functions $f(t)$ and $g(t)$, Equation (4.24) can be expressed below as

$$(\Delta_t \Delta_\omega)^2 = \left(\left[\int_{-\infty}^{\infty} t^2 |f(t)|^2 dt \right] \right) \left(\left[\int_{-\infty}^{\infty} g(t) \bar{g}(t) dt \right] \right) \qquad (4.25)$$

Recall that Equation (4.11) gives the following result for the first derivative of $f(t)$

$$\int_{-\infty}^{\infty} \left[\frac{df(t)}{dt} \right] e^{-i\omega t} dt = (i\omega) F(\omega) \qquad (4.26)$$

Applying Equation (4.26) to Equation (4.25) yields

$$(\Delta_t \Delta_\omega)^2 = \left(\left[\int_{-\infty}^{\infty} t^2 |f(t)|^2 dt \right] \right) \left(\left[\int_{-\infty}^{\infty} \left| \frac{df(t)}{dt} \right|^2 dt \right] \right) \qquad (4.27)$$

Since

$$\int_{-\infty}^{\infty} t\,f(t) \left[\frac{df(t)}{dt} \right] dt = \frac{1}{2} \int_{-\infty}^{\infty} t \left[\frac{df^2(t)}{dt} \right] dt$$

$$= \frac{t\,f^2(t)}{2} \bigg|_{-\infty}^{\infty} - \frac{1}{2} \int_{-\infty}^{\infty} f^2(t)dt \qquad (4.28)$$

$$= -\frac{1}{2}$$

Following from the Cauchy–Schwarz inequality [3], which states that

$$\left[\int_a^b \varphi_1(t)\varphi_2(t)dt \right]^2 \leq \left[\int_a^b [\varphi_1(t)]^2\,dt \right] \left[\int_a^b [\varphi_2(t)]^2\,dt \right] \qquad (4.29)$$

with $\varphi_1(t)$ and $\varphi_2(t)$ being two real integrable functions defined in $[a,\ b]$, Equation (4.27) becomes

$$(\Delta_t \Delta_\omega)^2 = \left(\left[\int_{-\infty}^{\infty} t^2\,|f(t)|^2 dt \right] \right) \left(\left[\int_{-\infty}^{\infty} \left| \frac{df(t)}{dt} \right|^2 dt \right] \right) \geq \left| \int_{-\infty}^{\infty} tf(t)\frac{df(t)}{dt}\,dt \right|^2 \qquad (4.30)$$

Substituting Equation (4.28) into Equation (4.30), we have $\Delta_t\,\Delta_\omega \geq \frac{1}{2}$, which is the uncertainty principle in Equation (4.21).

4.4   Short-Time Fourier Transform

The sinusoidal basis functions of the Fourier transform are global (infinite supported) and stationary with respect to time. They are inadequate for representing signals whose spectral responses are time varying. To address such a deficiency, a function $f(t)$ can be windowed using a proper window function $\varphi(t)$ that has desired time and frequency localizations. The Fourier transform of the windowed functions $f_\varphi(t) = f(t)\varphi(t - \tau)$ is then

$$F_\varphi(\tau,\ \omega) = \int_{-\infty}^{\infty} f_\varphi(t)\,e^{-i\omega t}\,dt$$

$$= \int_{-\infty}^{\infty} f(t)\,\varphi(t - \tau)\,e^{-i\omega t}\,dt \qquad (4.31)$$

$$= \langle f_\varphi(t),\quad e^{i\omega t} \rangle$$

$$= \langle f(t),\quad \bar{\varphi}(t - \tau)e^{i\omega t} \rangle$$

The window function $\varphi(t)$ is usually chosen to be compact and of short duration, thus enabling the Fourier transform of $f_\varphi(t) = f(t)\varphi(t - \tau)$ to reveal the signal's local spectral properties at each $t = \tau$ time location. The windowed Fourier transform defined above is therefore also referred to as the *short-time Fourier transform* (STFT).

As is suggested by the last equality in Equation (4.31), the windowed Fourier transform of $f(t)$ can be considered as if $f(t)$ is approximated using the basis functions $\varphi(t - \tau)e^{-i\omega t}$. By time-shifting $\varphi(t - \tau)$ through manipulating τ and by frequency-modulating $\varphi(t)$ using $e^{-i\omega t}$ at many frequencies ω, $f(t)$ is evaluated at every local (τ, ω), which is the tick mark in the simultaneous time-frequency plane.

For the inner products to be valid, $\varphi(t - \tau)e^{-i\omega t}$ have to be an orthonormal set and admissible. In addition, the Fourier transform of the window function, $\Phi(\omega)$, has to satisfy the condition

$$\Phi(\omega = 0) = \int_{-\infty}^{\infty} \varphi(t)dt \neq 0 \tag{4.32}$$

With a nonzero spectrum at $\omega = 0$, $\Phi(\omega)$ is essentially a low-pass filter. In other words, the requirement for knowing $f(t)$ for the entire time axis is greatly relaxed. Only the portion of $f(t)$ in the range where $\varphi(t - \tau)$ is defined is required. With a time duration Δ_t and a frequency bandwidth Δ_ω, the STFT $F_\varphi(\omega, \tau)$ defined in Equation (4.31) provides the temporal and spectral information about $f(t)$ in the region enclosed by $[\langle t \rangle + \tau - \Delta_t, \langle t \rangle + \tau + \Delta_t] \times [\langle \omega \rangle + \omega - \Delta_\omega, \langle \omega \rangle + \omega + \Delta_\omega]$.

The time function $f_\varphi(t) = f(t)\varphi(t - \tau)$ can be recovered through performing the following inverse Fourier transform:

$$\begin{aligned}
\frac{1}{2\pi} \int_{-\infty}^{\infty} F_\varphi(\tau, \omega)e^{i\omega k} d\omega &= \frac{1}{2\pi} \int_{-\infty}^{\infty} \left(\int_{-\infty}^{\infty} f(t)\varphi(t - \tau)e^{-i\omega t} dt \right) e^{i\omega k} d\omega \\
&= \frac{1}{2\pi} \int_{-\infty}^{\infty} \left(\int_{-\infty}^{\infty} f(t)\varphi(t - \tau)e^{i(k-t)\omega} dt \right) d\omega \\
&= \int_{-\infty}^{\infty} f(t)\varphi(t - \tau)\delta(k - t)dt \\
&= f(k)\varphi(k - \tau)
\end{aligned} \tag{4.33}$$

Replacing k with t, Equation (4.33) becomes

$$f(t) = \frac{1}{2\pi \varphi(0)} \int_{-\infty}^{\infty} F_\varphi(\tau, \omega)e^{i\omega t} d\omega \tag{4.34}$$

Thus $f(t)$ is obtained from $F_\varphi(\tau, \omega)$ following the formula in Equation (4.34).

The STFT integral in Equation (4.31) can be evaluated as a series sum, using digital computers, by properly sampling $f(t)$ and the window function $\varphi(t)$ as follows:

$$F_\varphi(\tau, \omega) \approx \Delta t \sum_{k=0}^{N-1} f(k\Delta t)\varphi(k\Delta t - \tau)e^{-i\left(\frac{2\pi n}{N\Delta t}\right)(k\Delta t)} \tag{4.35}$$

where Δt is the sampling rate, $k = 0, \ldots, N - 1$, and $n = -\frac{N}{2}, \ldots, \frac{N}{2}$. The discrete version of the STFT will be a good approximation of the corresponding integral transformation provided that a sufficient number of samples are taken for $f(t)$ and $\varphi(t)$ at a sampling rate that is equal to or higher than the corresponding Nyquist frequency.

4.5   Continuous-Time Wavelet Transform

In STFT, the time localization is provided by the window function while the frequency localization is established through the complex sinusoidal functions. Once the window function is selected, the associated time and frequency resolutions as determined by Δ_t and Δ_ω are fixed. They do not change from location to location in the time-frequency plane. Nor do they adapt to different or changing frequencies by adjusting themselves accordingly. More specifically, Δ_t does not increase its spread while resolving the low frequency contents, nor reduce its width while resolving the high frequency contents. This inherent rigidity can be relaxed by making the window function adaptive through scaling t, the time variable. Drawing inspiration from Equation (4.9), which states that scaling in the t-domain results in inverse scaling in the ω-domain, we see that if the center (mean) frequency of $\varphi(t)$ is ω_c, then the τ-translated, b-scaled version of the function, $\varphi\left(\frac{t-\tau}{b}\right)$, would have a time center at time τ and a scaled center frequency at $b\omega_c$ along with an equally scaled spectrum.

The continuous-time wavelet transform (CTWT or simply CWT) explores the scaling property above by employing orthonormal basis functions of particular attributes and features to provide adaptive time-frequency resolution. CTWT is a linear transformation mapping a finite energy function $f(t) \in L^2(-\infty, \infty)$ to the space spanned by the basis functions $\psi(t)$ as follows

$$
\begin{aligned}
W_\psi f(b, a) &= \int_{-\infty}^{\infty} f(t)\left[\frac{1}{\sqrt{a}}\bar{\psi}\left(\frac{t-b}{a}\right)\right] dt \\
&= \left\langle f(t),\ \frac{1}{\sqrt{a}}\psi\left(\frac{t-b}{a}\right)\right\rangle
\end{aligned}
\tag{4.36}
$$

where parameter b dictates the translation and $a > 0$ controls the dilation (scaling) of $\psi(t)$. The analyzing window function $\psi(t)$ has to have the following properties to be considered admissible

$$
\mathbf{\Psi}(\omega = 0) = \int_{-\infty}^{\infty} \psi(t)dt = 0
\tag{4.37}
$$

$$
\int_{-\infty}^{\infty} t^p\, \psi(t)dt = 0
\tag{4.38}
$$

Equation (4.37) indicates that $\psi(t)$ is finite-supported and oscillating like a wave, and is thus termed a wavelet. With $\mathbf{\Psi}(0) = 0$, that is, the Fourier transform of $\psi(t)$ vanishes at $\omega = 0$, the wavelet function also behaves like a band-pass filter. The second property in Equation (4.38) is an imposed constraint condition requiring that $\psi(t)$ has a higher p-order of vanishing moment, up to $p = 0, \ldots, M - 1$. The admissibility condition given in Equation (4.37) is traditionally stated using the following notations

$$
C_\psi = \frac{1}{2\pi}\int_{-\infty}^{\infty} \frac{|\mathbf{\Psi}(\omega)|^2}{|\omega|}d\omega < \infty
\tag{4.39}
$$

Equation (4.39) implies $\mathbf{\Psi}(0) = 0$, which is consistent with Equation (4.37).

Unlike the STFT where frequency information is provided by the complex sinusoidal functions, CTWT exploits the various versions of the dilated wavelet functions to establish simultaneous *time-scale* information. When the scaling parameter a in Equation (4.36) is reduced, the span (or support) of the wavelet $\psi(t)$ along the time axis is contracted, thereby allowing a larger range of frequencies to be resolved. However, when a is such that the corresponding time duration Δ_t is small, then the associated frequency bandwidth Δ_ω may be too large to differentiate low frequency components. Even though a implies frequency, and thus a measure of frequency, the reciprocal of a is in general not frequency.

Being characteristically a band-pass filter renders the wavelet function $\psi(t)$ to have two mean frequencies (centers) – one on each side of the axis at $\Psi(0) = 0$ in the Fourier domain. Since the two centers are symmetrical, and thus the mirror image of each other, it would suffice to consider only the positive frequencies to determine the frequency center and the width of the frequency spectrum. Following from the definitions given in Equations (4.18) and (4.20), we have the center and bandwidth of $\psi(t)$ defined using semi-infinite integrals [4]

$$\langle \omega \rangle_\psi^+ = \frac{\int_0^\infty \omega \, |\Psi(\omega)|^2 d\omega}{\int_0^\infty |\Psi(\omega)|^2 d\omega} \tag{4.40}$$

and

$$(\Delta_{\omega^+})_\psi^2 = \frac{\int_0^\infty (\omega - \langle \omega \rangle^+)^2 \, |\Psi(\omega)|^2 d\omega}{\int_0^\infty |\Psi(\omega)|^2 d\omega} \tag{4.41}$$

Note that $\int_0^\infty |\Psi(\omega)|^2 d\omega$ is half of the energy E defined according to Parseval's Theorem in Equation (4.15).

There is no similar condition of symmetry for the corresponding time center and time duration to apply to. As such their definitions remain the same as Equations (4.17) and (4.19),

$$\langle t \rangle_\psi = \frac{1}{\|\psi(t)\|^2} \int_{-\infty}^\infty t \, |\psi(t)|^2 dt \tag{4.42}$$

and

$$(\Delta_t)_\psi^2 = \frac{1}{\|\psi(t)\|^2} \left[\int_{-\infty}^\infty (t - \langle t \rangle_\psi)^2 \, |\psi(t)|^2 dt \right] \tag{4.43}$$

where the square of the norm of $\psi(t)$, $\|\psi(t)\|^2$ is by definition the energy E.

Although they have scaling-enabled adaptivity, wavelets are still subject to the abiding uncertainty principle. That is, without exception, the time-frequency resolution is bounded by

$$(\Delta_t)_\psi \, (\Delta_{\omega^+})_\psi \geq \frac{1}{2} \tag{4.44}$$

As a result, the wavelet transform $W_\psi f(b, a)$ in Equation (4.36) carries the local behavior of $f(t)$ analyzed by $\psi(t)$ in the time window

$$[a\langle t\rangle_\psi + b - a(\Delta_t)_\psi, \ a\langle t\rangle_\psi + b + a(\Delta_t)_\psi] \qquad (4.45)$$

and in the frequency window

$$\left[\frac{1}{a}(\langle\omega\rangle_\psi^+ - (\Delta_{\omega^+})_\psi), \ \frac{1}{a}(\langle\omega\rangle_\psi^+ + (\Delta_{\omega^+})_\psi)\right] \qquad (4.46)$$

Note that $2a(\Delta_t)_\psi \times \left(\frac{2}{a}\right)(\Delta_{\omega^+})_\psi = 4(\Delta_t)_\psi(\Delta_{\omega^+})_\psi$, which is the area of the time-frequency window. Thus, the area stays constant as expected.

With the wavelet function $\psi(t)$ satisfying the admissibility condition defined in Equations (4.37)–(4.39), the function $f(t)$ can be completely reconstructed by the sampled wavelet transform $W_\psi f(b, a)$ through the double-integral operation below

$$f(t) = \frac{1}{C_\psi} \int_{-\infty}^{\infty} \left\{ \int_0^{\infty} \frac{1}{a^2} [W_\psi f(b, a)]\overline{\psi}\left(\frac{t-b}{a}\right) da \right\} db \qquad (4.47)$$

As $\psi(t)$ is usually a real function, thus $\overline{\psi}(t) = \psi(t)$. Using the particular property, the equality below can be verified with ease

$$\int_{-\infty}^{\infty} |f(t)|^2 dt = \frac{1}{C_\psi} \int_{-\infty}^{\infty} \left\{ \int_0^{\infty} \frac{1}{a^2} |W_\psi f(b, a)|^2 da \right\} db \qquad (4.48)$$

which establishes that the weighted energy of $W_\psi f(b, a)$ in the wavelet time-scale plane equals the energy of the signal in the time domain. Equation (4.48) is generally considered the counterpart of Parseval's Theorem for the Fourier transform in Equation (4.15).

The discussion of the discrete version of CWT in Equation (4.36) using a dyadic scheme will be delayed until the chapter on wavelet filter banks as appropriate.

4.6   Instantaneous Frequency

As briefly discussed in Chapter 2, because the Fourier transform is mathematically linear and the spectral characteristics of a nonlinear, nonstationary signal vary with time, conventional Fourier-based analyses are inadequate in resolving the temporal progression of all individual spectral components in a signal. A new approach has to be developed to resolve the various predicaments identified and discussed previously. As an alternative to methods that are Fourier's in characteristic, the fundamental notion of instantaneous frequency is presented in this section, whose effectiveness in resolving the dilemmas is demonstrated using nonstationary responses obtained from several nonlinear dynamic systems.

4.6.1   Fundamental Notions

The frequency of a stationary signal is well defined following the Fourier approach. Generally, the frequency is defined as the number of oscillations per unit time of a physical field parameter such as displacement, current, or voltage. But for the nonstationary signals commonly encountered in communications, seismic, radar, medicine, sonar, and ultrasonic wave applications, this definition becomes ambiguous [5] and loses its effectiveness due to the fact that the spectral characteristics of the signals vary with time. Concepts that are viable and complete in describing this specific time-varying nature are thus needed. This is the practical motivation behind the notion of instantaneous frequency. Because instantaneous frequency is an intuitively sound and physically useful concept for describing signals of time-dependent spectral characteristics, efforts over the years have been focused on establishing a feasible definition for instantaneous frequency that is applicable to nonstationary signals of both monocomponent and multicomponent. Gabor [6] was the first to introduce a complex analytic signal, which was later employed to define instantaneous frequency as the time derivative of the phase of a signal by Ville [7]. The definition works well for monocomponent signals. However, it fails to produce physically reasonable results for multicomponent signals. Unfortunately, this failure not only hinders progress on developing a universally accepted definition of instantaneous frequency, but often leads to doubts and questioning of the necessity and existence of instantaneous frequency [8]. The fundamental concept of instantaneous frequency will be introduced in this section, and the problems with this definition will be analyzed using a few examples in sections that follow.

A time-varying signal $x(t)$ having both Amplitude Modulation (AM) and Frequency Modulation (FM) can be expressed as

$$x(t) = a(t)\cos(\phi(t)) \tag{4.49}$$

with $a(t)$ being the instantaneous amplitude and $\phi(t)$ the instantaneous phase. The corresponding complex form of the signal is then

$$z(t) = a(t)\exp(i\phi(t)) \tag{4.50}$$

where $z(t)$ is simply called the analytic signal. The time derivative of $\phi(t)$ is defined as the Instantaneous Frequency. However, there are infinite numbers of possible combinations of AM and FM for any given signal using this representation. The lack of uniqueness thus imparts no physical sense to the representation in Equation (4.50). To circumvent the problem and be able to obtain unique complex signals, Gabor [6] proposed an approach to "suppress the amplitudes belonging to negative frequencies and multiply the amplitudes of positive frequencies by two." Following this approach, Gabor's time domain complex signal, which is also an analytic signal by definition, can be determined using a slightly different definition as follows

$$\begin{aligned} z(t) &= x(t) + iy(t) \\ &= x(t) + iH[x(t)] \\ &= a(t)\exp(i\phi(t)) \end{aligned} \tag{4.51}$$

In Equation (4.51) $z(t)$, $a(t)$ and $\phi(t)$ are used in the same sense as before for convenience and $y(t)$ is the imaginary part of the analytic signal. $H[x(t)]$ is the Hilbert transform of the time-varying signal $x(t)$ defined as

$$y(t) = H[x(t)] = \frac{p}{\pi} \int_{-\infty}^{\infty} \frac{x(\tau)}{t - \tau} d\tau = x(t) * (p/_{\pi t}) \qquad (4.52)$$

where p is the Cauchy principle value. In theory $x(t)$ and $y(t)$ are out of phase by $\pi/2$. The instantaneous amplitude and phase are thus defined uniquely as

$$a(t) = \sqrt{x^2(t) + y^2(t)} \qquad (4.53)$$

$$\phi(t) = tan^{-1}(y(t)/x(t)) \qquad (4.54)$$

Building upon Equations (4.53) and (4.54), Ville [7] went further and defined the time derivative of the instantaneous phase as the instantaneous frequency

$$f(t) = \frac{1}{2\pi} \frac{d\phi(t)}{dt} = \frac{1}{2\pi} \frac{d}{dt}(tan^{-1}(y(t)/x(t))) \qquad (4.55)$$

The definition above captures the notion of instantaneity in nature and fits our intuitive expectation of the instantaneous frequency concept. It is encouraging that when the definition is applied to a sinusoidal signal, the obtained instantaneous frequency is exactly the frequency of the signal. As an illustration, Figure 4.1 shows the waveform of a sinusoidal signal and its corresponding instantaneous frequency determined using Equation (4.55). The frequency of this signal is initially 1 Hz and then abruptly switches to 3 Hz. The instantaneous frequency depicted in Figure 4.1(b) also shows this change. Figure 4.2 shows the waveform of a linear chirp signal, along with its corresponding instantaneous frequency and Fourier transform. The instantaneous frequency is seen to change from 1 Hz to 3 Hz linearly in Figure 4.2. This linear increase in frequency is not realized in the Fourier spectrum that also erratically displaying fictitious frequency components outside the 1–3 Hz range.

There are also other reasons for supporting this definition [3, 9]. From the following definition for the frequency spread associated with the complex analytic signal, $a(t) \exp(i\phi(t))$ [3],

$$\sigma_\omega^2 = \int_{-\infty}^{\infty} (a'(t))^2 dt + \int_{-\infty}^{\infty} (\phi'(t) - <\omega>^f)^2 a^2(t) dt \qquad (4.56)$$

where $<\cdot>^f$ is the average over the frequency domain. It is clear that the first integral represents the frequency spread from the Amplitude Modulation and the second term is due to the deviation of $\phi'(t)$ from the average frequency indicating explicitly that $\phi'(t)$ is closely related to the instantaneous frequency. In addition, the average frequency defined as follows,

$$<\omega>^f = \int_{-\infty}^{\infty} \omega |X(\omega)|^2 d\omega = \int_{-\infty}^{\infty} \phi'(t)|x(t)|^2 dt = <\phi'(t)>^t \qquad (4.57)$$

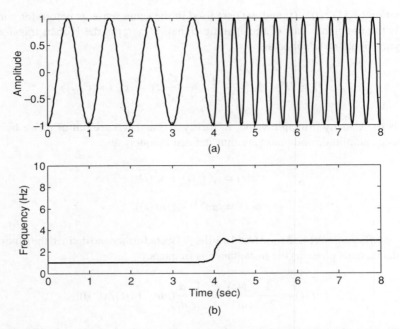

Figure 4.1 (a) Time history of a simple sinusoidal signal initially starting with 1 Hz and abruptly switching to 3 Hz. (b) Instantaneous frequency determined using the traditional definition

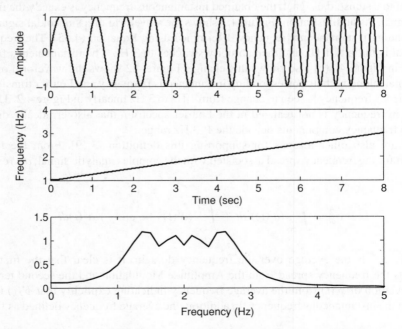

Figure 4.2 Linear chirp signal (top) and its corresponding instantaneous frequency (middle) and fast Fourier transform (bottom)

where $X(\omega)$ is the Fourier transform of $x(t)$ and $< \cdot >^t$ is the average over the time domain. Equation (4.57) suggests that the average frequency may be obtained by integrating the derivative of the phase with density over the time axis. It is thus considered appropriate and suitable to term the derivative of the phase the instantaneous frequency.

Because of these supporting reasons, the concept of instantaneous frequency has been intimately combined with the analytic signal and Hilbert Transform. However, the definition fails to work in cases involving multicomponent signals. This inevitably presents a constant difficulty for signal analysts because in the simple case of monocomponent signals the instantaneous frequency is often neither necessary nor required [9]. The dilemma between the seemly theoretically sound and physically appealing definition and the impracticality of the same definition to multicomponent signals has motivated a very long line of research aiming to remove the predicament of Ville's definition [3, 9, 10, 11, 12, 13, 14, 15, 16, 17, 18]. A historical review of the successive attempts to define instantaneous frequency was presented in [5]. The relationship between instantaneous frequency and analytic signal, group delay, and bandwidth-time product were also reviewed along with a thorough discussion on the relationship with time-frequency distribution. Cohen [3] also reviewed the development of Ville's definition and reported some doubts about the definition. Additionally, he studied the relationship of instantaneous frequency with group delay [10] and the ambiguity in the definition of instantaneous frequency and amplitude [11]. The proper use of Ville's definition, Cohen concluded, requires a choice to be made between the non-negative amplitude but an instantaneous frequency with infinite spikes, and a bounded instantaneous frequency but an instantaneous amplitude with negative values. Oliveira *et al.* [9] analyzed the advantages and shortcomings of Ville's definition and determined when and why the definition fails. Two alternative definitions were given by them: instantaneous frequency as an average frequency and instantaneous frequency as a heterodyning law. Loughlin *et al.* [12] proposed four conditions that the definition of instantaneous frequency should satisfy in order for the definition to have sound physical meaning. They also presented a method using the positive time-frequency distribution and time-varying coherent demodulation of the signal to find the instantaneous frequency. In the same vein, Barkat and Boashash [13, 14] recently demonstrated that the accuracy for estimating instantaneous frequency could be further enhanced using time-frequency distributions of better resolution.

4.6.2   Misinterpretation of Instantaneous Frequency

A few example signals are considered in order to analyze the reason for the failure of Ville's definition. Figure 4.3 presents an example two-component signal with equal amplitude

$$x(t) = a_0(\cos(\omega_1 t) + \cos(\omega_2 t)) \tag{4.58}$$

where a_0 is constant amplitude. Figure 4.4 also gives an example two-component signal but with unequal amplitude,

$$x(t) = a_1 \cos(\omega_1 t) + a_2 \cos(\omega_2 t) \tag{4.59}$$

where both a_1 and a_2 are constant amplitudes.

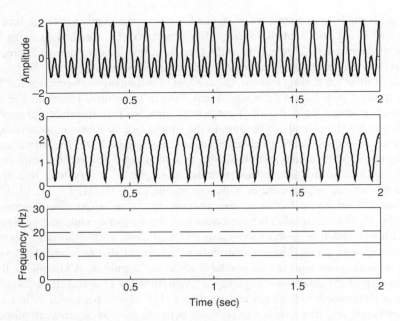

Figure 4.3 (a) A two-component signal with equal amplitude (b) Amplitude modulation of the signal (c) Instantaneous frequency determined using Ville's definition where dashed lines represent the two original frequencies

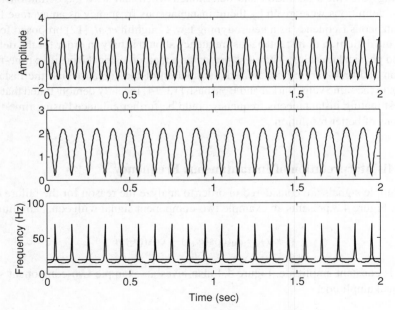

Figure 4.4 (a) A two-component signal with unequal amplitude (b) Amplitude modulation of the signal (c) Instantaneous frequency determined using Ville's definition, where dashed lines represent the two original frequencies

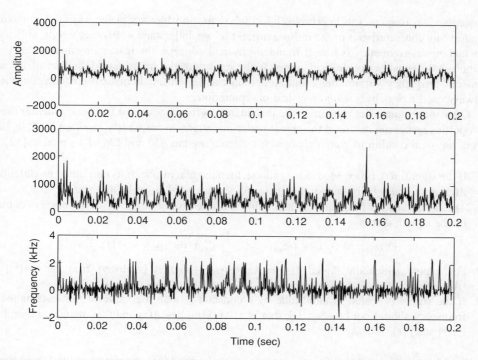

Figure 4.5 (a) A multicomponent signal (milling vibration data) (b) Amplitude modulation of the signal (c) Instantaneous frequency determined using Ville's definition. Reproduced with permission from [22]. Copyright 2004 Elsevier

The multicomponent signal shown in Figure 4.5 is a type of dynamic instability response that was recorded from a milling machine [19]. In all three figures, (a) plots the time history of the signal and (b) and (c), respectively, show their associated amplitude modulation and frequency modulation using Ville's definition. In Figure 4.3(c) and Figure 4.4(c), the dashed lines are the frequencies of the two components (ω_1 and ω_2) and the solid lines are the obtained instantaneous frequencies using Ville's definition.

Figure 4.4 is a typical example illustrating the various questions raised regarding the ambiguity of the definition of instantaneous frequency. Some of these are:

- The signal $x(t)$ consists of two constant amplitude and frequency signals. It is thus a stationary signal. However, its amplitude modulation and frequency modulation (instantaneous frequency) both demonstrate time-varying characteristics.
- There are only two frequencies, ω_1 and ω_2, in the signal but the obtained instantaneous frequency is severely obscured by becoming negative at times.
- Notwithstanding, why is there only one instantaneous frequency while the two frequencies ω_1 and ω_2 are physically present at all times?

Other questions raised regarding this definition can also be found in [3]. A meticulous examination of the amplitude and frequency modulations in Figure 4.5 concludes that the instantaneous frequency as interpreted by Ville's definition does not in any way conform to the underlying characteristics of the milling process as we understand it. In other words, although instantaneous frequency is a well-founded physical concept, the instantaneous frequency in Figure 4.4(c) makes no physical sense. Is the case with the traditional definition of instantaneous frequency yet another example demonstrating that mathematical elegance does not always ensure or deliver sound physical interpretation?

Questions about what properties the instantaneous frequency must have and whether these properties reveal the nature of the signal under investigation are widely reported [3, 5, 9, 12]. Four physical conditions were proposed for calculating the AM and FM of a signal in [12]:

1. If the signal, $x(t)$, is bounded in magnitude, then the magnitude of its amplitude modulation, $A(t)$, should also be bounded; that is, $|x(t)| < \infty$ implies $|A(t)| < \infty$.
2. The instantaneous frequency, $f(t)$, should be limited to the range of frequencies occupied by the signal; that is,

$$|X(\omega)|^2 = 0, \ \omega \notin (\omega_{\min} < \omega < \omega_{\max}) \ \Rightarrow \ \omega_{\min} < f(t) < \omega_{\max}$$

3. For a pure sinusoidal signal $x(t) = A_0 \cos(\omega_0 t + \phi_0)$, it is always true that $|A(t)| = |A_0|$ and $f(t) = \omega_0$.
4. If the signal is scaled in amplitude by a constant c, then the value of the instantaneous frequency should not be changed; that is, if $x(t)$ implies $A(t)$ and $f(t)$, then $cx(t)$ implies $cA(t)$ and $f(t)$.

Furthermore, for a multicomponent signal having individual constant amplitude and time-independent spectral contents, the instantaneous frequency should also be time independent; that is,

$$x(t) = \sum_{i=1}^{N} a_i e^{j(\omega_i t + \phi_i)} \quad implies \quad \frac{d}{dt} f(t) = 0 \tag{4.60}$$

The above condition was added at a later date by Oliverira and Barroso as the condition 5 in Ref. [9].

The signal in Figure 4.3 is also stationary. Its amplitude modulation is nonstationary, but its instantaneous frequency is both the average of the two frequencies and stationary, a case which has led to many researchers embracing Ville's definition. The average frequency at any one moment was even considered by some as one type of definition for instantaneous frequency [9, 12]. However, this perspective is not satisfactory. To demonstrate, consider a nonstationary signal defined below

$$x(t) = a_0(\cos(\omega_0 t + b t^2) + \cos(\omega_0 t - b t^2)) \tag{4.61}$$

where a_0 and b are constants. The signal is time dependent, but using Ville's definition the instantaneous frequency is determined to be ω_0. That is, the nonstationary signal was interpreted as having single stationary instantaneous frequency. Thus the frequency cannot tell us anything about the temporal evolution of all the frequencies in the signal. Therefore, it is unacceptable that the average frequency is defined as the instantaneous frequency even if the average frequency satisfies all the formerly listed five physical conditions. These five conditions are insufficient for completely defining instantaneous frequency.

It is obvious that the concept of instantaneous frequency defined following Ville's definition is in the average sense. Equations (4.56) and (4.57) are infinite integrals evaluated along time and thus are each an average in the infinite integral sense. The following definition using the time-frequency distribution also determines the average frequency at every time instance, instead of the true instantaneous frequency,

$$f(t) = \frac{\int_{-\infty}^{\infty} f\rho(t,\ f)\,df}{\int_{-\infty}^{\infty} \rho(t,\ f)\,df} \tag{4.62}$$

where $\rho(t,\ f)$ is the Fourier transform of the time-dependent autocorrelation function defined in [20]. There should be exactly two instantaneous frequencies, ω_1 and ω_2, in Figure 4.4, at each and every time moment. The averaging effect inherent in the definition not only averages the two frequencies in the stationary signal but also erroneously and erratically interprets the signal as nonstationary with negative instantaneous frequencies! It is beyond any argument that average frequency and instantaneous frequency are two different concepts. There can be only one average frequency at any one time, but many instantaneous frequencies. Ville's definition will work well for monocomponent signals and will understandably fall short on resolving multicomponent signals because it will make no discretion and henceforth interpret the average (in the infinite integral sense) of all presenting frequencies as the instantaneous frequency. Oliveira and Barroso [9] argued that " . . . the difficulties encountered in the previous example [i.e., Figure 4.4] are not intrinsic problems of traditional definition of Instantaneous Frequency, but instead a mere consequence of misuse of the concept." Because the studied signal has two components, and there is no point in asking what the instantaneous frequency of a multicomponent signal is. The rationale is that a multicomponent signal does not have one instantaneous frequency; by construction, it has several, at each moment. Furthermore, they considered that separating a signal into a set of different components remains an arbitrary human activity and, as such, the number of "components" can never be the key to the physical acceptability of the definition. These arguments are insightful but unfortunately misleading. The key to eliminating the various difficulties and ambiguities experienced in the previous examples ought to be finding a viable decomposition scheme. Seeking to develop or formulate another definition for instantaneous frequency from the notion of average frequency is missing the point. Boashash [5] also concluded the same, that Ville's definition has physical meaning only for monocomponent signals where there is only one frequency or a narrow range of frequencies varying as a function of time. For multicomponent signals, the notion of a single-valued instantaneous frequency becomes meaningless, and a breakdown into its components is required. Thus a decomposition method effective for decomposing a multicomponent signal into its associated monocomponent subsets and not obscuring or obliterating the physical essentials of the signal would allow the traditional definition of instantaneous frequency to be complete and applicable to signals of both mono- and multicomponent.

According to [21], it is indicated that the traditional definition of instantaneous frequency is not applicable to data that are not symmetric. Multicomponent signals are necessarily not symmetric but, by definition and also by requirement, a monocomponent subset is symmetric in time. Recall that the analytic signal in Equation (4.51), which has the form $a(t)\exp(i\phi(t))$, indicates that the physical meaning of the signal is implicitly carried by $a(t)$ and $\phi(t)$, the amplitude and frequency modulation, respectively. If the spectra of $a(t)$ and $\phi(t)$ are not separated in frequency, it can be shown that the Hilbert transform of the signal will be phase-distorted [5]. A multicomponent signal is generally broadband and more often than not the spectra of $a(t)$ and $\phi(t)$ will overlap. This is one of the reasons why the traditional definition of instantaneous frequency does not work for multicomponent signals. Because the Hilbert Transform-based analytical formulation provides better approximation to signals of relatively narrow bandwidth, the more closely a signal approaches the narrowband condition, the better in general the traditional definition will be in estimating the instantaneous frequency. Therefore, in order to find the true frequency progression within a multicomponent signal, it is necessary to break down the many, coupled components of the signal into its individual intrinsic components.

Thus, the instantaneous frequency for a multicomponent signal is defined as a global characteristic collectively described by the instantaneous frequencies associated with all the individual components in the signal, and the instantaneous frequency of each component can be uniquely determined using the traditional definition. The number of instantaneous frequencies is the same as the number of monocomponents in the signal. As an illustration, the signal $x(t) = \cos(\omega_1 t)\cos(\omega_2 t)$, can be decomposed into two distinct parts:

$$\frac{1}{2}\cos\left(\frac{\omega_1 + \omega_2}{2}\right) \quad \text{and} \quad \frac{1}{2}\cos\left(\frac{\omega_1 - \omega_2}{2}\right) \tag{4.63}$$

Applying Ville's definition to the individually decomposed parts and the two instantaneous frequencies, $\frac{\omega_1 + \omega_2}{2}$ and $\frac{\omega_2 - \omega_1}{2}$, within this signal would be readily resolved.

Contingent upon the availability of an effective decomposition scheme, Ville's definition would be a powerful tool allowing the temporal development of all frequency components to be represented by the corresponding instantaneous frequency at any time instance and thus enabling signals of monocomponent and multicomponent alike to be fully characterized. In the next section an effective decomposition method will be introduced and the feasibility of the scheme of decoupling a signal into its monocomponent subsets will be demonstrated using two of the examples previously considered in Figures 4.3(a) and 4.4(a).

4.6.3   Decomposition of Multi-Mode Structure

As stated, the traditional definition of instantaneous frequency is meaningful only for mono-component signals. In order to study the time-dependent characteristics of a multicomponent signal, the signal has to be decomposed into its monocomponent subsets, so that the instanta-neous frequency of all the intrinsic components can be subsequently determined following the traditional definition. To answer the question of how to extract these components and simulta-neously avoid turning the decomposition process into a mere "arbitrary" human activity, it is imperative that one provides sound answers to the following: How are the intrinsic components of a multicomponent signal defined and why do they individually have physical meanings?

It is appropriate to first examine what properties a physically meaningful component should have. From the perspective of amplitude and frequency modulation, a monocomponent signal should have one and only one instantaneous frequency at any time and its amplitude modulation and frequency modulation should be completely separated, meaning that their individual spectra are not overlapping. However, from the spectrum alone a signal cannot be discerned as monocomponent or multicomponent. As such, for all components, the associated analytic signal defined using the Hilbert transform is therefore needed. To ensure a good estimation of the instantaneous frequency using the Hilbert transform approach, the frequency modulation must be demodulated from the amplitude modulation. In other words, the spectra of $a(t)$ and $\phi(t)$ of the resultant analytic signal $a(t)\exp(i\phi(t))$ will need to be well separated. The relationship between the amplitude modulation and the frequency modulation is dependent upon the covariance defined as follows [3],

$$Cov_{t\,\omega} = <t\phi'(t)> - <t><\omega> \tag{4.64}$$

where

$$<t\phi'(t)> = \int t\phi'(t)|s(t)|^2 dt \tag{4.65}$$

$$<t> = \int t|s(t)|^2 dt \tag{4.66}$$

$$<\omega> = \int \omega|S(\omega)|^2 d\omega \tag{4.67}$$

and $<t\phi'(t)>$ is the average (again, in the infinite integral sense) of the product of time with instantaneous frequency. Since $<t\phi'(t)>$ equals $<t><\phi'(t)> = <t><\omega>$, if time and frequency are not mutually dependent then the covariance $Cov_{t\,\omega} = 0$. For a symmetric signal, the covariance between the amplitude and frequency modulation is 0 [3] implying that the amplitude and frequency modulations are completely demodulated. Hence symmetry seems desirable for the intrinsic component of a multicomponent signal.

An intrinsic component can be expressed as an analytic signal as $z_j(t) = a_j(t)\exp(i\phi_j(t))$, where the subscript j denotes the individual components so that $z(t) = \sum_{j=1}^{n} z_j(t)$. It should have only one frequency at any time instance and both its frequency and amplitude may change with time. Consider the following four possible physical scenarios for the component:

1. without amplitude and frequency modulation, the analytic component signal is a sinusoidal signal

$$z_j(t) = a_j \exp(i\omega_j t), \quad \text{where } a_j \text{ and } \omega_j \text{ are constants} \tag{4.68}$$

2. without amplitude modulation but with frequency modulation, the analytic component is

$$z_j(t) = a_j \exp(i\phi_j(t)), \quad \text{where } a_j \text{ is constant} \tag{4.69}$$

3. with amplitude modulation but without frequency modulation, the analytic component is

$$z_j(t) = a_j(t)\exp(i\omega_j t), \quad \text{where } \omega_j \text{ is constant} \tag{4.70}$$

4. with amplitude and frequency modulation, the analytic component signal can be written as

$$z_j(t) = a_j(t) \exp(i\phi_j(t)) \qquad (4.71)$$

For the simple cases (1) and (2), by definition, the real part of the analytic signal has to be symmetric in time. As long as the spectrum of $a_j(t)$ is well separated from (or much lower than) ω_j, the same can be concluded for case (3) that the real part of the analytic signal is symmetric. By the same token, it is reasonable to say that, if the spectrum of $a_j(t)$ is well separated from (much lower than) the spectrum of $\phi_j(t)$, the real part of the analytic signal in the complicated case (4) is also symmetric. It should be noted that by definition if the real part of the analytic signal is symmetric, then its corresponding imaginary part is also symmetric in time. Because the time-frequency covariance of a symmetric signal is 0, good separation of the spectra of $a_j(t)$ and $\phi_j(t)$ can always be maintained. In summary, satisfying the symmetric condition is thus a natural and meaningful requirement for the construction of the intrinsic components of a multicomponent signal.

Huang *et al.* [21] presented a decomposition scheme capable of breaking a multicomponent signal down into its intrinsic components. The method, referred to as the Empirical Mode Decomposition (EMD), was developed from the simple assumption that any signal consists of many simple intrinsic modes of oscillation. Each mode has the same numbers of extrema and zero-crossings, and the inherent oscillation is symmetric with respect to a local mean. The local mean is defined by the maximum envelope and minimum envelope without resorting to time scales. Given this local mean, modes of different time scales can be separated. Once separated, each mode is independent of each another and all modes will have no multiple extrema between successive zero-crossings. As such, each separated mode can be designated an Intrinsic Mode Function (IMF) by the following conditions: (a) the number of extrema and the number of zero-crossings must be either equal or different at most by one in the entire data set, and (b) the mean value of the envelope defined by the local maxima and the envelope defined by local minima is zero at every point. The two conditions fit perfectly our understanding of what is required of an intrinsic component.

An IMF represents a simple oscillatory mode. Given the two conditions required of an IMF, the first IMF can be extracted using the following steps:

Step 1: Identify and then connect all the local maxima of the signal $x(t)$ using a smooth function to obtain the maximum envelope function *max(t)*. Identify and connect all the local minima of $x(t)$ using a smooth function to obtain the minimum envelope function *min(t)*. Finally, set *mean1(t)=(max(t)+min(t))/2*.

Step 2: Subtract *mean1(t)* from the original signal, $x(t)$, so that *s1(t)= x(t) - mean1(t)*.

Step 3: Repeat Step 1 and 2 on *s1(t)* to obtain *mean2(t)* and *s2(t)* and afterwards, on *s2(t)* to obtain *mean3(t)* and *s3(t)*. Keep recursively repeating Step 1 and 2 until *sn(t)* is obtained that satisfies the two conditions of an IMF. Denote *C1(t)=sn(t)* and the difference between the original signal $x(t)$ and the first IMF as *d1(t)= x(t)- C1(t)*.

This sifting process is shown in Figure 4.6 using an example signal. Repeat the above procedures to get *C2(t)* and *d2(t)*, *C3(t)* and *d3(t)*,..., and *Cn(t)* and *dn(t)*, until *dn(t)* has less than two extrema or no IMF can be extracted from it. Denote *R(t)=dn(t)*, which is called the

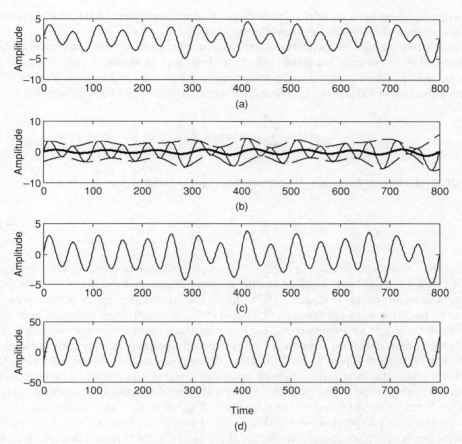

Figure 4.6 (a) A time domain signal. (b) Original signal (—) with its maximum envelope and minimum envelope (–·–·–) and its local mean (——). (c) $s1(t)$: result after one sifting process. (d) The first Intrinsic Mode Function obtained

residue. Thus the collection of all the IMFs $C1(t)$, $C2(t)$, ..., and $Cn(t)$ and the residue $R(t)$ restores back the original signal $x(t)$:

$$x(t) = \sum_{j=1}^{n} C_j(t) + R(t) \qquad (4.72)$$

The sifting process has two objectives: removing riding waves and making the signal profile more symmetric [21]. It can be seen from the following analysis that the first mode has the smallest time scale indicating that it includes the highest frequency components. As the decomposition goes on, the frequency components included in the IMF become lower. The residue should include almost no frequency components at all. Simply put, the decomposition is based on the local characteristic time scale of the data to produce an adaptive basis and it thus does not resort to a set of fixed time scales. Understandably the EMD is highly feasible for the analysis of nonlinear and nonstationary signals.

When all the intrinsic components are available, the corresponding monocomponent analytic signals can be constructed using the Hilbert transform and the obtained analytic signals can then be used to determine the signal's instantaneous frequencies and amplitude modulations using Equations (4.53) and (4.55). The original signal, $x(t)$, can now be expressed as

$$x(t) = \sum_{j=1}^{n} a_j(t) \exp(i \int \omega_j(t) dt) \qquad (4.73)$$

Alternatively, the signal can also be expressed using its Fourier representation as

$$x(t) = \sum_{j=1}^{\infty} a_j \exp(i\omega_j t) \qquad (4.74)$$

The Hilbert transform in Equation (4.52) can be appropriately interpreted as the convolution of $x(t)$ with $1/t$, thus clearly emphasizing the property of temporal locality of $x(t)$. The polar coordinate representation in Equation (4.71) further clarifies the local nature of the expression. It can be readily seen from Equations (4.73) and (4.74) that the Hilbert transform expression reveals the nature of an amplitude-varying and phase-varying trigonometric function of $x(t)$, while the Fourier transform expression does not show the property of amplitude variation.

It is evident from Equation (4.72) that the decomposition is mathematically complete. However, since the local mean acquired using the maximal and minimal envelopes is not exactly a mathematical mean, the IMFs obtained sequentially through the repetitive shifting process listed above are therefore not mutually orthogonal in the strict mathematical sense. Nevertheless, the two conditions of IMF, namely, symmetric in time and having no multiple extrema between zero-crossings, ensure that IMFs can be made almost orthogonal if certain conditions are met. It can be shown that the inner products of $c_i(t)$ with $c_j(t)$, and of $c_i(t)$ with $x(t) - c_j(t)$, where $i \neq j$, are almost zero. Therefore, although the local mean is not a true mean, the decomposition is both complete and orthogonal. In addition, Equation (4.72) shows that a multicomponent signal can be represented by its individual amplitude-varying and frequency-varying monocomponent whose amplitude modulation and frequency modulation are completely separated. The equation also makes it possible to represent the instantaneous frequency along with the corresponding amplitude as a function of time in a 3D-plot. For example, the amplitude can be plotted on the instantaneous time-frequency plane and this time-frequency distribution of amplitude is properly named as the Hilbert Amplitude Spectrum, $H(\omega, t)$, to note and differentiate from the more usual Fourier spectrum [22].

4.6.4   Example of Instantaneous Frequency

A demonstrative example is considered in this section. Figure 4.7(a) is a repetition of the same two-component signal as in Figure 4.4(a). Following the shifting procedures defined previously, the two intrinsic modes indicate a null signal. Note that the two obtained components C1 and C2 are the two sinusoids comprised of the signal. The traditional definition of instantaneous

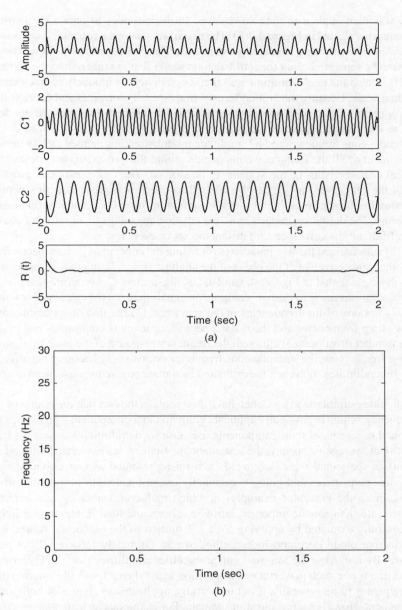

Figure 4.7 (a) A two-component signal with unequal amplitude and its two physical components obtained using the Empirical Mode Decomposition (b) The two instantaneous frequencies determined using the two components of the signal

frequency was then applied to the two resolved intrinsic modes to obtain the two instantaneous frequencies plotted in Figure 4.7(b). The two instantaneous frequencies are exactly what were expected without being averaged or obscured. The EMD enables multiple components (or modes) to be separated using the intrinsic time scales that are inherent to the signal (Equation (4.72)). It also fits one's intuition well that a component is uniquely characterized by its intrinsic time scale. Because the Hilbert transform temporally localizes and extracts the phase information pertaining to only one mode at a time, the phase velocity of the mode is then evaluated as the instantaneous frequency of the mode.

The instantaneous frequency and its amplitude modulation thus defined clearly demonstrate the development of all the frequency components within the two example systems and allow the physical characteristics of the systems to be studied. The four conditions given in [12], along with the one given in [9], can be applied to check and prove that this definition of instantaneous frequency on each and every separated intrinsic component surely satisfies all the conditions. The definition imparts concrete physical meanings to all intrinsic components and, henceforth, all the corresponding instantaneous frequencies!

Cohen [3] summarized the five paradoxes regarding the concept of instantaneous frequency defined as the time derivative of the phase of the analytic signal. Using two-component signals similar to those presented in Figures 4.3 and 4.4 as illustration, Cohen expressed his concerns over the observations made to the two-component signals that: (1) the calculated instantaneous frequency is not one of the frequencies in the spectrum; (2) the line of spectrum consists of only a few sharp frequencies and the instantaneous frequency is continuous and ranges over an infinite number of values; (3) although the frequency response of the analytic signal is zero for negative frequencies, the instantaneous frequency occasionally becomes negative; and (4) the signal is bandlimited, however, the evaluated instantaneous frequency is seen to go outside the band.

The difficulties summarized by Cohen have their roots in the fact that the notion of instantaneous frequency, which is physically applicable only to monocomponent signals, was inadvertently applied to signals of multicomponents. Because by definition instantaneous frequency is a conditional average frequency, the instantaneous frequencies associated with all inherent components of the signal were "averaged" and misrepresented as one obscured frequency with periodic sharp spikes and going occasionally out of bandwidth into the negative range as a result. From the presented examples in which multicomponent signals were properly decomposed into their monocomponent intrinsic subsets and then instantaneous frequencies were successfully extracted by applying Ville's definition to the individual subset, it is clear that an effective mode decomposition method would ultimately resolve all the paradoxes listed above. By removing the longstanding ambiguities and difficulties, the Empirical Mode Decomposition is one such powerful decomposition method enabling Ville's notion of instantaneous frequency to be generally, if not universally, applicable to signals of both mono- and multicomponent. However, a few questions remain. For example, can instantaneous frequency thus defined be related to the frequency defined in the Fourier sense? Or are they two totally different notions? Because the instantaneous frequency for a sinusoidal signal is the same as the frequency defined in the Fourier sense, the preferred idea is that the two concepts are related. If so, how exactly is the energy of the instantaneous frequency defined and how can it be related to the energy of the Fourier frequency? Huang, *et al.* [21] indicated that the energy of instantaneous frequency is related to the amplitude modulation. However, more specifics in this regard are needed.

4.6.5   Characteristics of Nonlinear Response

Bifurcation is the transition of a motion from one dynamic state to another accompanied by the appearance of new modes, the disappearance of old modes, or both. More specifically, a periodic motion contains only one frequency component (along with its harmonics), a quasi-periodic motion has at least two incommensurate frequencies (along with their harmonics), and a chaotic motion has a broadband spectrum. Thus, different states of periodic, period-doubling, quasi-periodic, or chaotic motions can be readily identified by their respective spectrum. This is one of the reasons that Fourier-based methods are often used in conjunction with time-domain methods for the identification of bifurcations. However, for the obvious arguments that the Fourier transform is linear and that its analyzing harmonic functions are stationary, spectral domain methods are not feasible for time-varying signals demonstrating nonlinearity. Because instantaneous frequency displays frequency variation with time, changes of dynamic states indicative of bifurcation and instability can be identified through monitoring the instantaneous frequencies and the corresponding amplitudes. By comparing the instantaneous frequencies of the old mode to that of the new mode, the type of bifurcation can be determined. For example, if the instantaneous frequency of a new mode is about half of the frequency of the old mode, period-doubling occurs. If the instantaneous frequency of the new mode is incommensurate with the old mode, quasi-periodic bifurcation occurs. Similarly, intermittence, crises, and chaotic motion can be thus determined.

   How does one quantify bifurcation and chaotic response of a dynamic nonlinear motion using instantaneous frequency? The idea is that a dynamic state can be determined simultaneously by observing the changes in time of the instantaneous frequency components and their corresponding energy. Because time information is essential for positively localizing aberrations, it is required that instantaneous frequency components occurring at the present moment be compared with those components that occurred one instance before. To prevent this qualitative comparison, an obviously enormous task, the cumulated weights of all instantaneous frequency components between a starting moment, t_0, and a selected ending moment, t_1, can be calculated as a function of the frequency as follows

$$h_{01}(\omega) = \int_{t_0}^{t_1} H(\omega, t)\, dt \qquad (4.75)$$

Thus $h_{12}(\omega)$, $h_{23}(\omega)$, ... can be similarly computed at moments between t_1 and t_2, t_2 and t_3, ..., respectively. The idea behind Equation (4.75) is borrowed from the Fourier transform in which the weights of all the harmonic components are evaluated over the whole time span. However, it should be noted that the marginal spectrum defined in Equation (4.75) describes and interprets the meaning of the frequency completely differently from the way the Fourier spectrum does. In the Fourier transform, presence of energy at a harmonic is interpreted as if the specific sinusoidal component is present throughout the entire duration of the time event. On the other hand, the marginal spectrum only gives the cumulated weights of all instantaneous frequency components over some selected time span in a probabilistic sense, thus indicating the occurrence probability of the frequency components being considered.

   If the motion is periodic before instant t_1, $h_{01}(\omega)$ will then have only one peak. If bifurcation occurs between t_1 and t_2, appearance of new modes will be registered as new frequency components in $h_{12}(\omega)$. This can be readily realized in the simultaneous instantaneous time-frequency plane as the secondary frequency component appears and its amplitude becomes larger. If dynamic stability is allowed to further deteriorate beyond t_2 to eventual chaotic states, more frequency components will exist in the marginal spectrum, $h_{23}(\omega)$, where broadband characteristics will be prominent. In addition, the amplitude variations of all the frequency components can be easily resolved either on the time-frequency plane or from the marginal spectrum at any instant t. The cumulated values of different frequency components can therefore be used to characterize a nonlinear system and to also quantify bifurcation and chaotic response of the system. In summary, the exact moment of bifurcation occurrence can be pinpointed using more refined marginal spectra. As an alternative demanding less computing requirement, the same task can be equally achieved in the instantaneous time-frequency domain.

In the immediately following an example is given to illustrate the use of instantaneous frequency to the detection and identification of bifurcation and nonlinear response. The example is a time-delay oscillator that displays pitchfork bifurcation and other types of bifurcation as results of nonlinear feedback. The time-delayed equation is given by

$$
\begin{aligned}
&\ddot{x} + \omega_0^2 x - \alpha_1 \dot{x} + \alpha_2 \dot{x}^3 \\
&= k \cos(\Omega t) + \beta_1 \left[\dot{x}(t - \tau) - \dot{x}(t) \right] + \beta_2 \left[\dot{x}(t - \tau) - \dot{x}(t) \right]^3
\end{aligned}
\tag{4.76}
$$

where system parameters α's, β's, and k are assumed to be positive and Ω is the external driving frequency. Given $\omega_0^2 = 10.0$, $\alpha_1 = 12.2$, $\alpha_2 = 0.3$, $\beta_1 = 6.5$, and $\Omega = 2.0$, different sets of periodic forcing (k's) and nonlinear feedback (β_2's) are studied in [23].

In the case presented in Figure 4.8 where β_2 is set to equal 0.1 and k is linearly increased from 1 to 10, a time step $\Delta t = 0.001$ sec is used to increment the response history, $x(t)$, and k and β_2 are each kept constant in the first 25 000 time steps. Starting at $t = 25$ sec the periodic forcing parameters are then linearly increased for the next 20 seconds (20 000 time steps) to their final values. Only the portion of the entire time response enclosing $t = 25$ sec and beyond is considered. As is evident from Figure 4.8(b), when k is relatively small, the driving frequency ($f = 1/\pi$) is not exactly prominent to begin with. The two IMFs resolved using EMD are seen in Figure 4.8(b) as two incommensurate modes demonstrating the attribute of a quasi-periodic motion. It is noted that only one of the modes in the figure displays temporal-modal structure with an oscillation period of approximately 0.4 sec. As no aberration is observed with either mode within the 19 sec window, no bifurcation or further change of dynamic state occurs for increasing the periodic forcing amplitude from 1.0 to 10.0. In addition, it can be shown that the accumulated effect from the forcing frequency becomes more prominent, while the effect from the natural frequency remains the same on the marginal spectrum, $h_{i,i+1}(\omega)$.

The concept of instantaneous frequency is of high practical significance for the analysis of nonlinear, nonstationary signals. Because their spectral characteristics vary with time, conventional Fourier-based analyses are inadequate in resolving the temporal progression of all individual spectral components in multicomponent signals. Ville's definition of instantaneous frequency works well for monocomponent signals; however, the definition falls short on providing a unified interpretation applicable also to multicomponent signals. The definition realizes infinite instantaneous frequency or negative amplitude in multicomponent signals; both of which are against our intuitive understanding of instantaneous frequency. As signals subject to daily investigation are mostly nonstationary and multicomponent in nature, it is essential

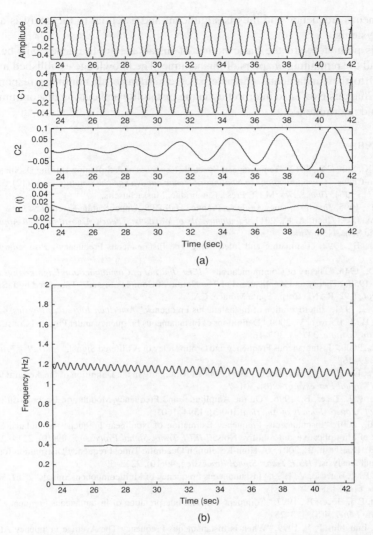

Figure 4.8 (a) Dynamic response and its decomposed intrinsic modes for $\beta_2 = 0.1$ and k being varied from 1.0 to 10.0 (b) Associated instantaneous frequencies corresponding to all resolved IMF C1-C2. Reproduced with permission from [23]. Copyright 2004 Elsevier

that the definition of instantaneous frequency be reexamined and physically established to enable wider application.

In this chapter, problems associated with the use of Ville's definition for instantaneous frequency were analyzed using a few simple examples, and a viable solution to the problems was identified. A decomposition method was employed to separate multicomponent signals into their associated monocomponent signal subsets. Ville's notion was then applied to the subsets to successfully extract the instantaneous frequencies inherent in the signal. The presented approach resolves the predicament commonly experienced in applying the definition of instantaneous frequency and thus is significant and essential to the analysis and study of nonlinear systems.

From this point forward the basic notion of instantaneous frequency will be the tool of choice for all the remaining chapters of this volume. The knowledge established in the chapter with regard to the fundamental characteristics common to all nonlinear responses will be applied in Chapters 6 to 11 to interpret dynamic instability and cutting discontinuities such as stick-slip and intermittence.

References

[1] Liu, M.-K., Suh, C.S., 2012, "Temporal and Spectral Responses of A Softening Duffing Oscillator Undergoing Route-To-Chaos," *Communications in Nonlinear Science and Numerical Simulations*, 17(12), 5217–28.

[2] Wiener, N., 1964, *Time Series*, M.I.T. Press, Cambridge, Massachusetts.

[3] Cohen, L., 1995, *Time-Frequency Analysis*, Prentice Hall PTR, Upper Saddle River, New Jersey.

[4] Goswami, J. C., Chan, A. K., 2011, *Fundamentals of Wavelets: Theory, Algorithms, and Applications*, John Wiley and Sons, New York.

[5] Boashash, B., 1992, "Estimating and Interpreting The Instantaneous Frequency," *Proceeding of IEEE*, 80, 520–68.

[6] Gabor, D., 1946, "Theory of Communications," *IEEE Journal of Communication Engineerings*, 93, 429–57.

[7] Ville, J., 1958, "Theory and Applications of The Notion of Complex Signal," Translated by I. Seline in *RAND Tech. Rept. T-92*, RAND Corp., Santa Monica, CA.

[8] Mandel, L., 1974, "Interoperation of Instantaneous Frequency," *American Journal of Physics*, 42(10), 840–46.

[9] Oliveira, P. M., Barroso, V., 2000, "Definitions of Instantaneous Frequency under Physical Constraints," *Journal of Franklin Institute*, 337(4), 303–16.

[10] Cohen, L., 2000, "Instantaneous Frequency and Group Delay of A Filtered Signal," *Journal of Franklin Institute*, 337(4), 329–46.

[11] Cohen, L., Loughlin, P., Vakman, O., 1999, "On an Ambiguity in the Definition of the Amplitude and Phase of a Signal," *Signal Processing*, 79(3), 301–7.

[12] Loughlin, P. J., Tacer, B., 1996, "On the Amplitude- and Frequency-Modulation Decomposition of Signals," *Journal of Acoustic Society of America*, 100(3), 1594–1601.

[13] Barkat, B., 2001, "Instantaneous Frequency Estimation of Nonlinear Frequency-Modulated Signals in The Presence of Multiplicative and Additive Noise," *IEEE Trans. Signal Processing*, 49(10), 2214–22.

[14] Barkat, B., Boashash, B., 2001, "A High-Resolution Quadratic Time–Frequency Distribution for Multicomponent Signal Analysis," *IEEE Trans. Signal Processing*, 49(10), 2232–39.

[15] Oliveira, P.M., Barroso, V., 1999, "Instantaneous Frequency of Multicomponent Signals" *IEEE Signal Processing Letter*, 6(4) 81–83.

[16] Loughlin, P. J., Tacer, B., 1997, "Comments On The Interpretation of Instantaneous Frequency," *IEEE Signal Processing Letter*, 4(5), 123–25.

[17] Nho, W., Loughlin, P. J., 1999, "When Is Instantaneous Frequency The Average Frequency At Each Time?" *IEEE Signal Processing Letter*, 6(4), 78–80.

[18] Wei, D., Bovick, A. C., 1998, "On The Instantaneous Frequencies of Multicomponent AM-FM Signals," *IEEE Signal Processing Letter*, 5(4), 84–86.

[19] Suh, C.S., Khurjekar, P.P., Yang, B., 2002, "Characterization and Identification of Dynamic Instability in Milling Operation," *Mechanical Systems and Signal Processing*, 15(5), 829–48.

[20] L. Cohen, 1996, "Generalized Phase-space Distribution," *Journal of Mathematical Physics*, 7(5), 781–86.

[21] Huang, N.E., Shen, Z., Long, S. R., *et al.*, 1998, "The Empirical Mode Decomposition and Hilbert Spectrum for Nonlinear and Nonstationary Time Series Analysis," *Proceedings of Royal Society, London Series A*, 454(1971), 903–95.

[22] Yang, B., Suh, C. S., 2003, "Interpretation of Crack-Induced Rotor Nonlinear Response Using Instantaneous Frequency," *Mechanical Systems and Signal Processing*, 18(3), 491–513.

[23] Yang, B. and Suh, C.S., 2004 "On The Nonlinear Features of Time-Delayed Feedback Oscillators," *Communications in Nonlinear Science and Numerical Simulations*, 9(5), 515–29.

5

Wavelet Filter Banks

The concept of discrete wavelet transform (DWT) is central to the nonlinear time-frequency control to be developed in Chapter 7. In the overall architecture of the time-frequency control, the DWT decomposition serves to represent a dynamic response with different levels of spectral resolution without losing the corresponding temporal information. Any anomaly indicative of perturbation or instability is identified and properly addressed in the wavelet domain. The conditioning of the response is realized by an adaptive filtered-x LMS algorithm that updates the wavelet representations of the response. These representations are then synthesized to create a conditioned response as the control output that meets a specific control target. Having a fundamental knowledge of the working principle and implementation of DWT is therefore essential for developing a full comprehension for the nonlinear time-frequency control. As one of the physical components of the time-frequency control, the DWT decomposition algorithm and the corresponding synthesis algorithm incorporates a popular dyadic scheme [1]. The various basic properties of DWT and IDWT render it possible for them to be implemented as digital filter banks for fast computations. These properties are best understood, and the underlying essences best presented, using the fundamental notions of multiresolution analysis (MRA). A concise, while also relatively comprehensive, discussion of wavelets is given in the chapter. We will start with an illustrative example using the Haar wavelet.

5.1 A Wavelet Example

A function $f(t) \in L^2(-\infty, \infty)$ can be completely represented by the wavelet basis function $\psi(t)$ using a dyadic scheme. By replacing the dilation and translation parameters with $a = 2^{-m}$ and $b = n2^{-m}$, respectively, where m, n are integers, in Equation (4.36), a popular version of the discrete wavelet transform emerges below

$$
\begin{aligned}
d_{m,n} &= W_\psi f(n2^{-m}, 2^{-m}) \\
&= 2^{m/2} \int_{-\infty}^{\infty} f(t)\bar{\psi}(2^m t - n)dt \\
&= \int_{-\infty}^{\infty} f(t)\bar{\psi}_{m,n}(t)dt
\end{aligned}
\tag{5.1}
$$

Control of Cutting Vibration and Machining Instability: A Time-Frequency Approach for Precision, Micro and Nano Machining, First Edition. C. Steve Suh and Meng-Kun Liu.
© 2013 John Wiley & Sons, Ltd. Published 2013 by John Wiley & Sons, Ltd.

where the dilated and translated wavelet functions are denoted as follows

$$\psi_{m,n}(t) = 2^{m/2}\bar{\psi}(2^m t - n) \tag{5.2}$$

Equation (5.1) implies that a set of orthogonal functions exists, $\hat{\psi}_{m,n}(t)$, that are the dual functions of $\psi_{m,n}(t)$, that satisfy these conditions below:

$$\langle \psi_{m,n}(t), \hat{\psi}_{k,l}(t) \rangle = \delta_{m,k}\delta_{n,l} \tag{5.3}$$

where $\delta_{m,k}$ is the Kronecker delta, defined as

$$\delta_{m,k} = \begin{cases} 1, & m = k \\ 0, & m \neq k \end{cases} \tag{5.4}$$

and

$$d_{m,n} = \langle f(t), \psi_{m,n}(t) \rangle = \langle f(t), \hat{\psi}_{m,n}(t) \rangle \tag{5.5}$$

In the case that $\hat{\psi}_{m,n}(t) = \psi_{m,n}(t)$, which necessarily implies that $\psi_{m,n}(t)$ are orthonormal, the piecewise approximation of $f(t)$ is complete

$$f(t) = \sum_{m=-\infty}^{\infty} \sum_{n=-\infty}^{\infty} d_{m,n}\psi_{m,n}(t) \tag{5.6}$$

To establish a proper interpretation of Equation (5.2), which is essential for the proper representation of a function $g(t) \in L^2(\mathbf{R})$ using the discrete wavelet transform, we consider the simple piecewise constant function $\phi(t)$ defined as

$$\phi(t) = \begin{cases} 1, & 0 \leq t < 1 \\ 0, & \text{otherwise} \end{cases} \tag{5.7}$$

$\phi(t)$ is an orthonormal set because

$$\int \phi(t - m)\bar{\phi}(t - n)dt = \delta_{m,n} \tag{5.8}$$

In addition, the scaled and translated versions of $\phi(t)$ also satisfy the condition of orthogonality,

$$\int \phi_{m,n}(t)\bar{\phi}_{k,l}(t)dt = \int \{2^{m/2}\phi(2^m t - n)\}\{2^{k/2}\bar{\phi}(2^k t - l)\}dt \\ = \delta_{m,k}\,\delta_{n,l} \tag{5.9}$$

Equations (5.9) and (5.6) imply that $g(t)$ can be approximated by at least the following two different versions of the basis function $\phi(t)$:

$$g(t) \approx \sum_{n=-\infty}^{\infty} c_{0,n} \phi_{0,n}(t) = \sum_{n=-\infty}^{\infty} c_{0,n} \phi(2^0 t - n) \tag{5.10}$$

and

$$g(t) \approx \sum_{n=-\infty}^{\infty} c_{1,n} \phi_{1,n}(t) = 2^{1/2} \sum_{n=-\infty}^{\infty} c_{1,n} \phi(2^1 t - n) \tag{5.11}$$

where $c_{0,n}$ and $c_{1,n}$ are the coefficients associated with scales $m = 0$ and $m = 1$, respectively. Since the support of $\phi(2t - n)$ is only half of $\phi(t - n)$, of the two approximations of $g(t)$, Equation (5.11) will suffer from less approximation error. In other words, $\phi(2t - n)$ provides better approximation resolution than $\phi(t - n)$. This is further supported by the fact that

$$\phi(t) = \phi(2t) + \phi(2t - 1) \tag{5.12}$$

which states that the low resolution basis $\phi(t)$ can be completely represented by its scaled version, $\phi(2t)$, the high resolution basis function. Therefore if the space defined by $2^{1/2} \phi(2^1 t - n)$ is denoted as V_1, then in terms of approximation resolution, V_0 the space defined by $2^0 \phi(2^0 t - n)$, is a subset of V_1. That is,

$$V_0 \subset V_1 \tag{5.13}$$

Equations (5.12) and (5.13) necessarily imply that

$$\phi(2t) = c_0 \phi(t) + d_0 \psi(t) \tag{5.14}$$

where c_0 and d_0 are constants, and $\psi(t)$ is a function that provides all the detailed information not carried by $\phi(t)$. The $\psi(t)$ corresponding to the piecewise function $\phi(t)$ in Equation (5.7) that satisfies Equation (5.14) is the infamous Haar wavelet function defined as [2]

$$\psi(t) = \begin{cases} 1, & 0 \leq t < \dfrac{1}{2} \\ -1, & \dfrac{1}{2} \leq t < 1 \\ 0, & \text{otherwise} \end{cases} \tag{5.15}$$

The piecewise function $\psi(t)$ is also an orthonormal set having the following properties similar to $\phi(t)$,

$$\int \psi(t - m) \bar{\psi}(t - n) dt = \delta_{m,n} \tag{5.16}$$

$$\int \psi_{m,n}(t) \bar{\psi}_{k,l}(t) dt = \int \{2^{m/2} \psi(2^m t - n)\} \{2^{k/2} \bar{\psi}(2^k t - l)\} dt$$
$$= \delta_{m,k} \delta_{n,l} \tag{5.17}$$

By satisfying the condition below, $\psi(t)$ and $\phi(t)$ are mutually orthogonal:

$$\int \phi(t-m)\bar{\psi}(t-n)dt = \delta_{m,n} \tag{5.18}$$

In addition,

$$\phi(2t) = \frac{1}{2}\phi(t) + \frac{1}{2}\psi(t) \tag{5.19}$$

and

$$\phi(2t-1) = \frac{1}{2}\phi(t) - \frac{1}{2}\psi(t) \tag{5.20}$$

Drawing an analogy from Equations (5.19) and (5.20), it can be shown that the approximation of $g(t)$ in Equation (5.11) can be expressed by using the n-translated version of Equation (5.14) as

$$g(t) \approx 2^{1/2} \sum_{n=-\infty}^{\infty} c_{1,n}\phi(2^1 t - n)$$

$$= 2^{1/2} \sum_{n=-\infty}^{\infty} \left[\left(\frac{c_{1,2n} + c_{1,2n+1}}{2}\right)\phi(t-n) + \left(\frac{c_{1,2n} - c_{1,2n+1}}{2}\right)\psi(t-n)\right] \tag{5.21}$$

Therefore if the space spanned by $2^0\psi(2^0 t - n)$ is denoted as W_0, then in terms of approximation resolution and the associated detail, Equation (5.21) indicates that W_0 is an orthogonal complementary space of V_0 in V_1, denoted as

$$V_1 = V_0 \oplus W_0 \tag{5.22}$$

where W_0 is also a subset of V_1, that is,

$$W_0 \subset V_1 \tag{5.23}$$

5.2   Multiresolution Analysis

From the illustrative example above we see that the basis function $2^{m/2}\phi(2^m t - n)$ defines the V_m space which is a subspace of $L^2(\mathbf{R})$. For any $g(t) \in V_m$, the following is true,

$$g(t) = \sum_{m=-\infty}^{\infty}\sum_{n=-\infty}^{\infty} c_{m,n}\phi_{m,n}(t) = 2^{\frac{m}{2}}\sum_{n=-\infty}^{\infty} c_{m,n}\phi(2^m t - n) \tag{5.24}$$

The $\phi(t)$ that satisfies Equation (5.24) is a *scaling function*. Equation (5.24) can be generalized to cover all possible integer scales $m \in \mathbf{Z}$ so that

$$\{0\} \cdots \subset V_{-2} \subset V_{-1} \subset V_0 \subset V_1 \subset V_2 \cdots L^2(\mathbf{R}) \tag{5.25}$$

and

$$\text{if} \quad g(t) \in V_m \quad \text{then} \quad g(2t) \in V_{m+1} \tag{5.26}$$

We see that if $\phi(t)$ defines V_0, then $\phi(t) \in V_1$. $\phi(t)$ is also in the space defined by $\phi(2t)$. As such, $\phi(t)$ can be approximated by $\phi(2t)$, the scaled versions of itself, as

$$\phi(t) = \sum_{n=-\infty}^{\infty} h_n \, 2^{\frac{1}{2}} \phi(2t - n) \tag{5.27}$$

with h_n being the corresponding scaling function coefficients. Equation (5.27) is termed a recursive dilation equation. The multiresolution analysis (MRA) equation is another popular name for it. In the case of the piecewise approximation in Equation (5.12), $h_0 = h_1 = \frac{1}{\sqrt{2}}$.

To explore the frequency representations of the scaling function $\phi(t)$ at two consecutive scales, Fourier transform is applied to Equation (5.27) below

$$\int_{-\infty}^{\infty} \phi(t) e^{-i\omega t} dt = 2^{\frac{1}{2}} \sum_n h_n \left[\int_{-\infty}^{\infty} \phi(2t - n) e^{-i\omega t} dt \right] \tag{5.28}$$

A closer examination of Equation (5.28) reveals that

$$\int_{-\infty}^{\infty} \phi(t) e^{-i\omega t} dt = \Phi(\omega) \tag{5.29}$$

and, after performing a simple change of variable, the term on the right becomes

$$\begin{aligned}
2^{\frac{1}{2}} \sum_{n=-\infty}^{\infty} h_n \left[\int_{-\infty}^{\infty} \phi(2t - n) e^{-i\omega t} dt \right] &= \frac{\sqrt{2}}{2} \left(\sum_{n=-\infty}^{\infty} h_n e^{-i(\frac{\omega}{2})n} \right) \left[\int_{-\infty}^{\infty} \phi(\tau) e^{-i(\frac{\omega}{2})\tau} d\tau \right] \\
&= \frac{1}{\sqrt{2}} \mathbf{H}\left(\frac{\omega}{2}\right) \Phi\left(\frac{\omega}{2}\right)
\end{aligned} \tag{5.30}$$

where $\Phi(\omega)$ is the Fourier transform of the scaling function $\phi(t)$ and $\mathbf{H}\left(\frac{\omega}{2}\right)$ is the discrete Fourier transform of the set of scaling coefficients h_n. Because $\sum_{n=-\infty}^{\infty} h_n e^{-i(\frac{\omega}{2})n}$ is a convolution and $\mathbf{H}(0) \neq 0$, $\mathbf{H}\left(\frac{\omega}{2}\right)$ is therefore the frequency response of a low-pass type filter. From Equations (5.28)–(5.30), the recursive equation in Equation (5.27) has the following spectral equivalent:

$$\begin{aligned}
\Phi(\omega) &= \frac{1}{\sqrt{2}} \mathbf{H}\left(\frac{\omega}{2}\right) \Phi\left(\frac{\omega}{2}\right) \\
&= \left[\frac{1}{\sqrt{2}} \mathbf{H}\left(\frac{\omega}{2}\right) \right] \left[\frac{1}{\sqrt{2}} \mathbf{H}\left(\frac{\omega}{4}\right) \Phi\left(\frac{\omega}{4}\right) \right] \\
&= \prod_{j=1}^{\infty} \left[\frac{1}{\sqrt{2}} \mathbf{H}\left(\frac{\omega}{2^j}\right) \right] \Phi(0)
\end{aligned} \tag{5.31}$$

Provided that $\Phi(0)$ is well-defined, Equation (5.31) indicates that the scaling function $\phi(t)$ is characterized by the low-pass filter $\mathbf{H}(\omega)$. That is, $\phi(t)$ can be generated using the low-pass filter $\mathbf{H}(\omega)$.

Next we consider an important property of the scaling function that is also related to the derivation of wavelet functions. Given that $\phi(t)$ is an orthonormal set of basis functions,

$$\int_{-\infty}^{\infty} \phi(t)\bar{\phi}(t-n)dt = \delta_{0,n} \tag{5.32}$$

and by Parseval's Theorem, the corresponding energy in the spectral domain is therefore

$$\int_{-\infty}^{\infty} \Phi(\omega)\bar{\Phi}(\omega)e^{-i\omega n}d\omega = 2\pi\,\delta_{0,n} \tag{5.33}$$

The sum of the integer n taken on both sides of Equation (5.33) yields

$$\sum_{n=-\infty}^{\infty} [2\pi\,\delta_{0,n}] = 2\pi \tag{5.34}$$

and

$$\sum_{n=-\infty}^{\infty} \left[\int_{-\infty}^{\infty} \Phi(\omega)\bar{\Phi}(\omega)e^{-i\omega n}d\omega \right] = \int_{-\infty}^{\infty} \Phi(\omega)\bar{\Phi}(\omega) \sum_{n=-\infty}^{\infty} [e^{-i\omega n}]d\omega$$

$$= 2\pi \int_{-\infty}^{\infty} \Phi(\omega)\bar{\Phi}(\omega) \sum_{n=-\infty}^{\infty} [\delta(\omega - 2n\pi)]d\omega \tag{5.35}$$

Equating the last equation to Equation (5.34) we have

$$\sum_{m=-\infty}^{\infty} |\Phi(\omega + 2m\pi)|^2 = 1 \tag{5.36}$$

Equation (5.22) can be generalized to represent V_2 using the three consecutive orthogonal complementary spaces of coarser resolution as follows

$$\begin{aligned} V_2 &= V_1 \oplus W_1 \\ &= V_0 \oplus W_0 \oplus W_1 \\ &= V_{-1} \oplus W_{-1} \oplus W_0 \oplus W_1 \end{aligned} \tag{5.37}$$

Following from the nesting defined in Equation (5.25), it is concluded that for scaling approaching negative infinity, that is, for $m \to -\infty$, Equation (5.37) becomes

$$L^2(\mathbf{R}) = \cdots W_{-2} \oplus W_{-1} \oplus W_0 \oplus W_1 \oplus \cdots \tag{5.38}$$

Since

$$W_{-\infty} \oplus \cdots W_{-2} \oplus W_{-2} \oplus W_{-1} \oplus W_0 = V_1 \qquad (5.39)$$

or in a more compact, while also general, form

$$\overset{k-1}{\underset{j=-\infty}{\oplus}} W_j = V_k \qquad (5.40)$$

the *differences* between the spaces defined by the scaling functions of lower (coarser) scales can be arbitrarily chosen to define the scaling space of the higher scale. If $\psi(t)$ spans the difference, or *detail*, space W_0, and $W_0 \subset V_1$, then $\psi(t)$ can be represented by the basis scaling functions that span the space V_1 as

$$\psi(t) = \sum_{n=-\infty}^{\infty} g_n 2^{\frac{1}{2}} \phi(2t - n) \qquad (5.41)$$

where g_n is the corresponding coefficient. Because W_j is an orthogonal complementary space of V_j and $V_j \subset V_{j+1}$, it can be shown through applying the condition of orthogonality defined in Equation (5.18) that

$$g_n = (-1)^n h_{1-n} \qquad (5.42)$$

From the derivations above we see that $2^{m/2}\psi(2^m t - n)$, the dilations and translations of $\psi(t)$, form an orthonormal basis that defines the V_m space which is a subspace of $L^2(\mathbf{R})$, thus for any $g(t)$

$$g(t) = \sum_{m=-\infty}^{\infty} \sum_{n=-\infty}^{\infty} d_{m,n} \psi_{m,n}(t) = 2^{\frac{m}{2}} \sum_{n=-\infty}^{\infty} d_{m,n} \psi(2^m t - n) \qquad (5.43)$$

where $d_{m,n}$ is determined through the following inner product operation

$$d_{m,n} = \int_{-\infty}^{\infty} g(t) \bar{\psi}_{m,n}(t) dt \qquad (5.44)$$

The function $\psi(t)$ that satisfies Equation (5.43) is a wavelet function and the constant $d_{m,n}$ defined in Equation (5.44) is a wavelet series coefficient.

The nestings in Equations (5.25) and (5.38) signify that $L^2(\mathbf{R})$ can be spanned by a set of functions $\phi_{m,n}(t)$ and $\psi_{m,n}(t)$. Since $V_j = V_{j-1} \oplus W_{j-1}$ and

$$L^2(\mathbf{R}) = V_{j-1} \oplus W_{j-1} \oplus W_j \oplus \cdots \qquad (5.45)$$

for any function $f(t) \in L^2(\mathbf{R})$ that is defined below which is represented by the j-scale scaling functions in V_j

$$f(t) = \sum_{n=-\infty}^{\infty} c_{j,n} \phi_{j,n}(t) \tag{5.46}$$

or equivalently

$$f(t) = \sum_{n=-\infty}^{\infty} c_{l,n} \phi_{l,n}(t) + \sum_{m=l}^{j-1} \sum_{n=-\infty}^{\infty} d_{m,n} \psi_{m,n}(t) \tag{5.47}$$

Equation (5.47) is an expansion of $f(t)$ by the scaling and wavelet functions. In other words, $f(t)$ can be represented as the linear combination of $\phi_{l,n}(t)$ and $\psi_{m,n}(t)$.

Recalling Equation (5.36) and making use of the result from the first half of Equation (5.31) we have

$$\sum_{m=-\infty}^{\infty} |\Phi(\omega + 2m\pi)|^2 = \sum_{m=-\infty}^{\infty} \left| \mathbf{H}\left(\frac{\omega + 2m\pi}{2}\right) \Phi\left(\frac{\omega + 2m\pi}{2}\right) \right|^2 = 4 \tag{5.48}$$

By considering separately the odd and even numbers of the summation index m, Equation (5.48) takes up a slightly different form below

$$\sum_{m=-\infty}^{\infty} \left| \mathbf{H}\left(\frac{\omega + 2m\pi}{2}\right) \Phi\left(\frac{\omega + 2m\pi}{2}\right) \right|^2$$

$$= \sum_{l=-\infty}^{\infty} \left| \mathbf{H}\left(\frac{\omega + l\pi}{2}\right) \Phi\left(\frac{\omega + l\pi}{2}\right) \right|^2 \tag{5.49}$$

$$+ \sum_{l=-\infty}^{\infty} \left| \mathbf{H}\left[\frac{\omega + (2l+1)(2\pi)}{2}\right] \Phi\left[\frac{\omega + (2l+1)(2\pi)}{2}\right] \right|^2$$

As the frequency response of h_n, $\mathbf{H}(\omega) = \mathbf{H}(\omega + 2\pi)$, that is, $\mathbf{H}(\omega)$ is 2π periodic, Equation (5.49) can be manipulated further below

$$\sum_{l=-\infty}^{\infty} \left| \mathbf{H}\left(\frac{\omega + l\pi}{2}\right) \Phi\left(\frac{\omega + l\pi}{2}\right) \right|^2 + \sum_{l=-\infty}^{\infty} \left| \mathbf{H}\left[\frac{\omega + (2l+1)(2\pi)}{2}\right] \Phi\left[\frac{\omega + (2l+1)(2\pi)}{2}\right] \right|^2$$

$$= \left| \mathbf{H}\left(\frac{\omega}{2}\right) \right|^2 \sum_{l=-\infty}^{\infty} \left| \Phi\left(\frac{\omega + l\pi}{2}\right) \right|^2 + \left| \mathbf{H}\left(\frac{\omega + 2\pi}{2}\right) \right|^2 \sum_{l=-\infty}^{\infty} \left| \Phi\left(\frac{\omega + (2l+1)(2\pi)}{2}\right) \right|^2$$

$$= 4 \tag{5.50}$$

As a result, Equation (5.50) can be reduced to an identity equation as follows

$$|\mathbf{H}(\omega)|^2 + |\mathbf{H}(\omega + \pi)|^2 = 1 \tag{5.51}$$

or in terms of their complex conjugates as

$$\mathbf{H}(\omega)\bar{\mathbf{H}}(\omega) + \mathbf{H}(\omega + \pi)\bar{\mathbf{H}}(\omega + \pi) = 1 \tag{5.52}$$

It can be shown that there exists a function $\mathbf{G}(\omega) = -e^{-i\omega}\bar{\mathbf{H}}(\omega + \pi)$ that forms a quadrature mirror filter pair [3] with $\mathbf{H}(\omega)$ and satisfies the condition below:

$$\mathbf{H}(\omega)\bar{\mathbf{G}}(\omega) + \mathbf{H}(\omega + \pi)\bar{\mathbf{G}}(\omega + \pi) = 0 \tag{5.53}$$

Since $\mathbf{G}(0) = 0$ and $\mathbf{G}(\pi) = 1$, thus $\mathbf{G}(\omega)$ is a high-pass filter.

Readers can verify that the wavelet function $\psi(t)$ defined in Equation (5.41) has the following property:

$$\mathbf{\Psi}(\omega) = \mathbf{G}\left(\frac{\omega}{2}\right)\mathbf{\Phi}\left(\frac{\omega}{2}\right) = \mathbf{G}\left(\frac{\omega}{2}\right)\prod_{j=2}^{\infty}\left[\frac{1}{\sqrt{2}}\mathbf{H}\left(\frac{\omega}{2^j}\right)\right] \tag{5.54}$$

in which $\mathbf{\Psi}(\omega)$ is the Fourier transform of $\psi(t)$ and the result in Equation (5.31) is applied. Thus $\mathbf{G}(\omega)$ is the frequency response of the coefficient g_n in Equation (5.41) that satisfies the condition with h_n in Equation (5.42).

Equations (5.27) and (5.41) are referred to as the two-scale relations. Using the two coefficient sequences $\{h_n, g_n\} \in L^2(\mathbf{R})$, whose relation is well defined in Equation (5.42), one can generate the two sets of orthogonal basis functions $\phi(t)$ and $\psi(t)$ that are fundamental to defining the MRA (multiresolution analysis) space. This can be further elucidated by considering the subspace V_{j+1} and the two orthogonal complementary subspaces of coarser resolution, V_j and W_j. Their relationship is defined by

$$\phi(2^j t) = \sum_{n=-\infty}^{\infty} h_n \phi(2^{j+1}t - n) \tag{5.55}$$

$$\psi(2^j t) = \sum_{n=-\infty}^{\infty} g_n \psi(2^{j+1}t - n) \tag{5.56}$$

Thus the two-scale relations are a basic notion applicable to the entire MRA space. The Fourier transforms of Equations (5.55) and (5.56) are

$$\mathbf{\Phi}(\omega) = \left[\frac{1}{2}\sum_{n=-\infty}^{\infty} h_n e^{\frac{-i\omega n}{2}}\right]\mathbf{\Phi}\left(\frac{\omega}{2}\right) = \mathbf{H}\left(e^{\frac{-i\omega n}{2}}\right)\mathbf{\Phi}\left(\frac{\omega}{2}\right) \tag{5.57}$$

$$\mathbf{\Psi}(\omega) = \left[\frac{1}{2}\sum_{n=-\infty}^{\infty} g_n e^{\frac{-i\omega n}{2}}\right]\mathbf{\Phi}\left(\frac{\omega}{2}\right) = \mathbf{G}\left(e^{\frac{-i\omega n}{2}}\right)\mathbf{\Phi}\left(\frac{\omega}{2}\right) \tag{5.58}$$

Following the same recursive scheme by which Equations (5.31) and (5.54) were obtained, Equations (5.57) and (5.58) can be expressed as

$$\Phi(\omega) = \prod_{j=1}^{\infty} \mathbf{H}\left(e^{-i\frac{\omega}{2^j}}\right) \tag{5.59}$$

$$\Psi(\omega) = \mathbf{G}\left(e^{-i\frac{\omega}{2}}\right) \prod_{j=2}^{\infty} \mathbf{H}\left(e^{-i\frac{\omega}{2^j}}\right) \tag{5.60}$$

Equations (5.59) and (5.60) are the general forms of Equations (5.31) and (5.54), respectively.

As an illustration, Figure 5.1 shows the Daubechies-3 (db3) scaling functions and corresponding wavelets generated using the two general recursive equations in Equations (5.55) and (5.56) and the coupling equation in Equation (5.41). The orthogonal coefficients h_n and g_n followed are tabulated in Table 5.1. The coefficient pairs $\{h_n, g_n\}$ for the Daubechies wavelet family up to db10 can be found in many sources and references on the wavelet theory [4], including Wikipedia at http://en.wikipedia.org/wiki/Daubechies_wavelet.

With the two-scale relations developed, the decomposition relation between two consecutive scales can be derived. For simplicity we will consider the case of $V_1 = V_0 + W_0$. Since $\phi(2t)$ and $\phi(2t - 1)$ are both in V_1, Equation (5.47) along with Equations (5.55) and (5.56) indicates that

$$\phi(2t) = \sum_{n=-\infty}^{\infty} [h_{2n}\phi(t - n) + g_{2n}\,\psi(t - n)] \tag{5.61}$$

$$\phi(2t - 1) = \sum_{n=-\infty}^{\infty} [h_{2n-1}\phi(t - n) + g_{2n-1}\,\psi(t - n)] \tag{5.62}$$

The equations above can be generalized into the following

$$\phi(2^{j+1}t - k) = \sum_{n=-\infty}^{\infty} [h_{2n-k}\,\phi(2^j t - n) + g_{2n-k}\,\psi(2^j t - n)] \tag{5.63}$$

The infinite series in Equation (5.63) is therefore a mathematical description of $V_{j+1} = V_j + W_j$ that allows the scaling function at an arbitrary scale to be represented by the scaling and wavelet functions at one scale lower. It is also called the synthesis equation for obvious reasons.

Table 5.1 Daubechies-3 (db3) low-pass and high-pass filter coefficients h_n and g_n

$h_1 = 0.33267055295095688$	$g_1 = h_6$
$h_2 = 0.80689150931333875$	$g_2 = -h_5$
$h_3 = 0.45987750211933132$	$g_3 = h_4$
$h_4 = -0.13501102001039084$	$g_4 = -h_3$
$h_5 = -0.085441273882241486$	$g_5 = h_2$
$h_6 = 0.035226291882100656$	$g_6 = -h_1$

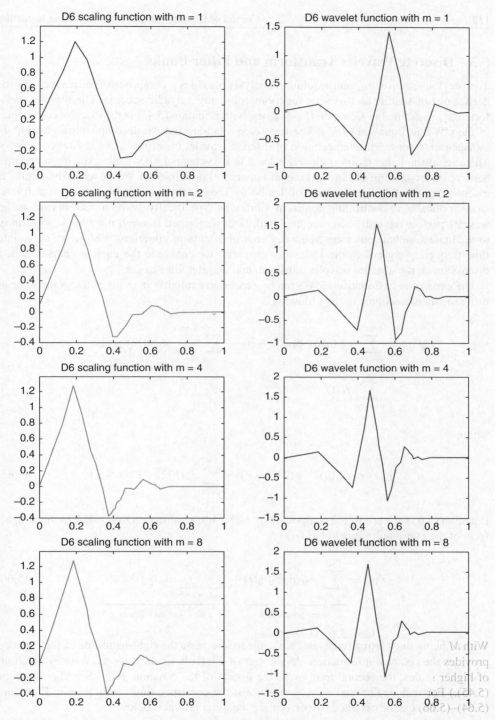

Figure 5.1 Daubechies-3 (db3) scaling functions (left column) and corresponding wavelet functions (right column) at 4 different dyadic scales m = 1, 2, 4, and 8 (from top to bottom)

5.3   Discrete Wavelet Transform and Filter Banks

The previous section on multiresolution analysis provides a comprehensive treatment of the derivations of scaling and wavelet functions following a dyadic scheme. The wavelet transform $d_{m,n}$ as defined in Equation (5.44) along with Equation (5.43) is the discrete counterpart of the CWT in Equation (4.36). The two-scale relations and the decomposition relation are indispensable for the implementation of discrete wavelet transform using digital computers. This digital implementation is closely related to two-channel filter banks. Having a comprehensive understanding of the working and operation principles of the wavelet filter banks is essential for meeting the objective of this book. That is, to establish the fundamental framework applicable to controlling nonlinear vibrations. Specifically, wavelet decomposition and wavelet perfect reconstruction are explored, both performed through filter bank operations, to facilitate simultaneous time-frequency control of cutting vibrations. Before we delve into time-frequency control in the following chapters, we conclude the current chapter with a discussion on the discrete wavelet transform and wavelet filter banks.

The expansion in Equation (5.47) can be expressed explicitly in terms of a specific scale M using the dyadic scheme in the following

$$f(t) = \sum_{n=-\infty}^{\infty} c_M(n) 2^{\frac{M}{2}} \phi(2^M t - n) + \sum_{m=M}^{\infty} \sum_{n=-\infty}^{\infty} d_m(n) 2^{\frac{m}{2}} \psi(2^m t - n)$$

$$= \sum_{n=-\infty}^{\infty} f_n(t) \tag{5.64}$$

where

$$f_n(t) = c_M(n) 2^{\frac{M}{2}} \phi(2^M t - n) + \sum_{m=M}^{\infty} d_m(n) 2^{\frac{m}{2}} \psi(2^m t - n) \tag{5.65}$$

Equation (5.65) represents the component of $f(t)$ at each translation interval. A more compact form of the expansion Equation (5.64) is

$$f(t) = \underbrace{\sum_{n=-\infty}^{\infty} c_M(n) \phi_{M,n}(t)}_{\text{coarse information}} + \underbrace{\sum_{m=M}^{\infty} \sum_{n=-\infty}^{\infty} d_m(n) \psi_{m,n}(t)}_{\text{high resolution detail}} \tag{5.66}$$

With M being the starting (coarsest) scale, the first term on the right-hand side of the equation provides the coarsest information. As the rest of $L^2(\mathbf{R})$ is spanned by the wavelet functions of higher scales, the second term gives the details of the function $f(t)$. (See also Equation (5.45).) For real scaling and wavelet functions, the corresponding coefficients in Equations (5.64)–(5.66) can be obtained by performing the inner products below

$$c_M(n) = \int_{-\infty}^{\infty} f(t) \phi_{M,n}(t) dt = \langle f(t), \phi_{M,n}(t) \rangle \tag{5.67}$$

and

$$d_m(n) = \int_{-\infty}^{\infty} f(t)\psi_{m,n}(t)\,dt = \langle f(t), \psi_{m,n}(t) \rangle \qquad (5.68)$$

Equations (5.67) and (5.68) are the *discrete wavelet transform* (DWT) of the function $f(t)$[5]. Coefficients $c_M(n)$ and $d_m(n)$ completely describe $f(t)$ in a way mathematically similar to Fourier series coefficients, though of very different physical interpretation and implication. The orthonormal properties of the scaling and wavelet functions assure that the energy of the function $f(t)$ is conserved in the Parseval sense as follows:

$$\int_{-\infty}^{\infty} |f(t)|^2\,dt = \sum_{n=-\infty}^{\infty} |c_M(n)|^2 + \sum_{m=M}^{\infty} \sum_{n=-\infty}^{\infty} |d_m(n)|^2 \qquad (5.69)$$

Instead of the frequency components, it is the time translation n and the dyadic scale M that together define the energy in the transform domain.

The recursive equation that corresponds to subspace V_1 spanned by $\phi(2t - n)$ is

$$\phi(t) = \sum_{n=-\infty}^{\infty} h_n \sqrt{2}\, \phi(2t - n) \qquad (5.70)$$

Then at scale j and i translation step, the corresponding scaling function is

$$\phi(2^j t - i) = \sum_{n=-\infty}^{\infty} h_n \sqrt{2}\, \phi(2^{j+1} t - 2i - n) = \sum_{k=-\infty}^{\infty} h_{(k-2i)}\sqrt{2}\, \phi(2^{j+1} t - k) \quad (5.71)$$

For function $f(t) \in L^2(\mathbf{R})$ defined in V_{j+1} (see also Equation (5.46))

$$f(t) = \sum_{n=-\infty}^{\infty} c_{j+1,n}\phi_{j+1,n}(t)$$

$$= \sum_{n=-\infty}^{\infty} c_{j+1,n} 2^{\frac{j+1}{2}} \phi(2^{j+1} t - n) \qquad (5.72)$$

To represent $f(t)$ at a lower scale j, wavelet functions are required as they provide the detailed information not available in the coarse representation. Thus we have

$$f(t) = \sum_{n=-\infty}^{\infty} c_{j,n} 2^{\frac{j}{2}} \phi(2^j t - n) + \sum_{n=-\infty}^{\infty} d_{j,n} 2^{\frac{j}{2}} \psi(2^j t - n) \qquad (5.73)$$

Simple inner product operations can be followed to determine the two coefficients

$$c_{j,n} = \langle f(t), \phi_{j,n}(t) \rangle = \int_{-\infty}^{\infty} f(t) 2^{\frac{j}{2}} \phi(2^j t - n)\,dt \qquad (5.74)$$

and

$$d_{j,n} = \langle f(t), \psi_{j,n}(t) \rangle = \int_{-\infty}^{\infty} f(t) 2^{\frac{j}{2}} \psi(2^j t - n) \, dt \qquad (5.75)$$

Incorporating Equation (5.71) into Equation (5.74), followed by a little manipulation, one has

$$c_{j,n} = \sum_{k=-\infty}^{\infty} h_{(k-2n)} \left(\int_{-\infty}^{\infty} f(t) 2^{\frac{j+1}{2}} \phi(2^{j+1} t - k) \, dt \right)$$

$$= \sum_{k=-\infty}^{\infty} h_{(k-2n)} (c_{j+1,k}) \qquad (5.76)$$

The reader can show that the wavelet coefficients have a similar relationship with the one scale higher scaling coefficients:

$$d_{j,n} = \sum_{k=-\infty}^{\infty} g_{(k-2n)} (c_{j+1,k}) \qquad (5.77)$$

Equations (5.76) and (5.77) explicitly state that $c_{j,n}$ and $d_{j,n}$, the discrete wavelet transform of $f(t)$, can be *simultaneously* computed by a low-pass filter (of h_n impulse response) and a high-pass filter (of g_n impulse response) that run *in parallel*. Such a filter arrangement is called a filter bank [3, 6]. Through the filter bank operation the wavelet transform is performed without having to compute the wavelet function $\psi(t)$.

The filtering of $c_{j+1,k}$ to generate $c_{j,n}$ and $d_{j,n}$ is facilitated through convolving the expansion coefficients $c_{j+1,k}$ with h_n and g_n in time. As is signified by the index $(k - 2n)$, the convolutions defined in Equations (5.76) and (5.77) are to be performed at time instance k, not on the entire data, but rather on a reduced set of the data. Specifically the reduced set is generated by taking every other data point, thus omitting half of the information carried by the original data set. This is physically implemented by convolving $c_{j+1,k}$ by the time-reversed recursive impulse responses h_{-n} and g_{-n}, followed by decimating or down-sampling by 2. This process of parallel filtering and down-sampling is repeated on the scaling coefficients following the two-channel configuration shown in Figure 5.2. The three decomposition levels illustrated in the figure are the embodiment of the following MRA analysis:

$$V_{j+1} = V_{j-2} \oplus W_{j-2} \oplus W_{j-1} \oplus W_j \qquad (5.78)$$

which is an operation going from the fine scale to coarse scale.

Discrete wavelet transform also has a counterpart through which the decomposed data can be reconstructed without error. This particular counterpart is a unique discrete inverse transform of the DWT that satisfies certain conditions to render perfect recovery of the original data from its components at different scales. Let us consider a signal $f_{j+1}(t)$ that is defined in the V_{j+1} space. The multiresolution analysis ensures that $f_{j+1}(t)$ can be represented by the approximation of itself $f_j(t)$ defined in a one scale lower space in V_j along with the detail

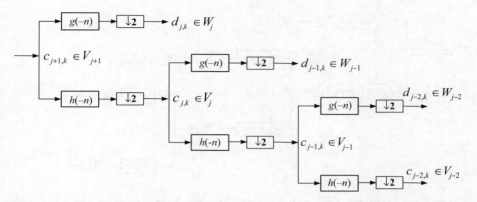

Figure 5.2 DWT performed by two-channel analysis filter banks showing three levels of decomposition

$g_j(t)$ defined in the complementary space W_j. The mathematical statement of the above is therefore

$$f_{j+1}(t) = f_j(t) + g_j(t)$$
$$= \sum_{n=-\infty}^{\infty} c_{j,n}\phi_{j,n}(t) + \sum_{n=-\infty}^{\infty} d_{j,n}\psi_{j,n}(t) \tag{5.79}$$

By incorporating the two-scale relations in Equations (5.27) and (5.41), the synthesis operation in Equation (5.79) can be expressed in terms of the impulse responses and scaling functions at two consecutive scales as follows

$$\sum_{k=-\infty}^{\infty} c_{j+1,k}\phi(2^{j+1}t - k) = \sum_{n=-\infty}^{\infty} c_{j,n} \sum_{k=-\infty}^{\infty} h_k^*\phi(2^{j+1}t - 2n - k)$$
$$+ \sum_{n=-\infty}^{\infty} d_{j,n} \sum_{k=-\infty}^{\infty} g_k^*\phi(2^{j+1}t - 2n - k) \tag{5.80}$$

The right-hand side can be rearranged to become

$$\sum_{n=-\infty}^{\infty} c_{j,n} \sum_{k=-\infty}^{\infty} h_k^*\phi(2^{j+1}t - 2n - k) + \sum_{n=-\infty}^{\infty} d_{j,n} \sum_{k=-\infty}^{\infty} g_k\phi(2^{j+1}t - 2n - k)$$
$$= \sum_{n=-\infty}^{\infty} c_{j,n} \sum_{M=-\infty}^{\infty} h_{(M-2n)}^*\phi(2^{j+1}t - M) + \sum_{n=-\infty}^{\infty} d_{j,n} \sum_{M=-\infty}^{\infty} g_{(M-2n)}^*\phi(2^{j+1}t - M) \tag{5.81}$$
$$= \sum_{M=-\infty}^{\infty} \left(\sum_{n=-\infty}^{\infty} h_{(M-2n)}^*c_{j,n} + \sum_{n=-\infty}^{\infty} g_{(M-2n)}^*d_{j,n} \right)\phi(2^{j+1}t - M)$$

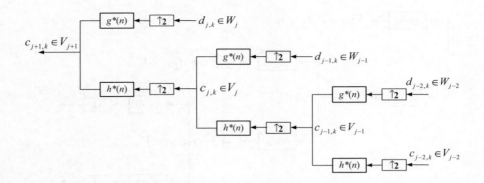

Figure 5.3 Inverse DWT performed by two-channel synthesis filter banks showing three levels of reconstruction

Since the choice of summation index is arbitrary, it can be concluded from comparing the two equations that

$$c_{j+1,k} = \sum_{n=-\infty}^{\infty} h^*_{(k-2n)}(c_{j,n}) + \sum_{n=-\infty}^{\infty} g^*_{(k-2n)}(d_{j,n}) \qquad (5.82)$$

Equation (5.82) is the counterpart of Equations (5.76) and (5.77), representing the inverse transform operation that restores a signal from scaling functions and wavelet coefficients of a coarser scale. Note that a different set of notions were used in deriving the inverse transform above. Impulse responses with a superscript star are understood to be associated with the reconstruction algorithm. They are related to the decomposition filters as follows: $h^*_n = h_n$ and $g^*_n = g_n$.

The recovery of $c_{j+1,k}$ from $c_{j,n}$ and $d_{j,n}$ is also realized through convolutions. Unlike the decomposition algorithm in Equations (5.76) and (5.77), however, the index $(k - 2n)$ in the reconstruction equation requires the convolutions to be performed after an up-sampling process. Up-sampling is done by inserting zeros between each data point. Thus the process serves to double the length of the j-scale scaling function and wavelet coefficient sequences before they reach the filter bank. The combining algorithm is implemented in Figure 5.3.

References

[1] Chui, C.K., 1992, *An Introduction to Wavelets*, Academic Press, San Diego, CA.
[2] Rioul, O., Vetterli, M., 1991, "Wavelet and Signal Processing," *IEEE Signal Processing Magazine*, 8(1) 14–38.
[3] Strang, G., Nguyen, T., 1996, *Wavelets and Filter Banks*, Wellesley-Cambridge Press, Wellesley, MA.
[4] Daubechies, I., 1992, *Ten Lectures on Wavelets*, SIAM, Philadelphia, PA.
[5] Goswami, J. C., Chan, A. K., 2011, *Fundamentals of Wavelets: Theory, Algorithms, and Applications*, John Wiley and Sons, New York.
[6] Mallat, S.G., 2008, *A Wavelet Tour of Signal Processing*, 3rd edn, Academic Press, Burlington, MA.

6

Temporal and Spectral Characteristics of Dynamic Instability

It was shown earlier that turning is a transient process that is inherently nonlinear and complex. Unlike the dynamic system studied in Chapter 4, whose response is time-invariant and stationary, real-world cutting processes in general are nonstationary with a time-varying spectrum. To be able to control cutting with stability it is imperative that the basic characteristics of the process be understood and the underlying parameters be identified. In the present chapter a popular nonlinear system is explored to establish its fundamental property in response to nonstationarity. The conclusions drawn are generally applicable to all the dynamic systems considered in the following chapters.

One of the essential objectives in studying a nonlinear system is to obtain the condition that guarantees the existence of periodic solutions so that their stabilities can be subsequently determined [1]. Steady-state solution is obtained for small but finite amplitude oscillations around the equilibrium point to estimate the threshold value of the excitation amplitude, stability region, and number of limit cycles. Linearization is performed under the assumption that if the operation range is in the immediate proximity of the equilibrium point of the nonlinear system, the response of the linearized model would approximate the nonlinear one with accuracy. However, there are cases where, although giving a correct time profile of the nonlinear response, the inherent components resolved using perturbation methods neither collectively nor individually provide any physically meaningful representation of the nonlinear system [2]. Applying linearization to investigate nonlinear systems without exercising proper discretion will obscure the underlying nonlinear characteristics and risk misinterpreting the stability bound.

Fourier-based analyses have been widely accepted as a tool for exploring nonlinear systems. Because stationary sinusoids are employed in representing time-varying signals of inherent nonlinearity, the use of Fourier domain methodologies will also risk misrepresenting the underlying physics of the nonlinear system being investigated [3]. As

Control of Cutting Vibration and Machining Instability: A Time-Frequency Approach for Precision, Micro and Nano Machining, First Edition. C. Steve Suh and Meng-Kun Liu.
© 2013 John Wiley & Sons, Ltd. Published 2013 by John Wiley & Sons, Ltd.

most methods employed to process nonstationary signals are Fourier-based, they also suffer from the shortcomings associated with Fourier transform [4]. The fact that nonlinear responses, including route-to-chaos, are intrinsically transient and nonstationary with coupled amplitude-frequency modulation, implies that, if a nonlinear response is to be fully characterized, the inherent amplitude modulation (AM) and frequency modulation (FM) need to be temporally decoupled [4]. The concept of instantaneous frequency (IF) introduced in Chapter 4 is adopted to resolve the dependency of frequency on time. Growing attention is focused on the Hilbert–Huang transform (HHT), which has been used to investigate the response of quadratic and cubic nonlinearities [5], Duffing oscillators [6], dynamic systems with slowly-varying amplitudes and phases [7], and fault induced nonlinear rotary [8]. Because HHT does not use predetermined basis functions and their orthogonality for component extraction, it provides an instantaneous amplitude and frequency of the extracted components for the accurate estimation of system characteristics and nonlinearities [9]. It is shown that HHT is more appropriate than sinusoidal harmonics for characterizing nonstationary and transient responses. The interpretation of nonlinearity using IF is found to be both intuitively rigorous and physically valid.

Various Duffing oscillators have been explored to help elucidate a wide range of physical applications in the real-world. In [10] the response of a damped Duffing oscillator with harmonic excitation is analyzed by second-order perturbation solutions along with Floquet analysis to predict symmetry-breaking and period-doubling bifurcation. Duffing oscillators under nonstationary excitations are also considered by many, where linear and cyclic variations of frequencies and amplitudes are applied and nonstationary bifurcation is studied. It is shown that the nonstationary process is distinct from the stationary process with different characteristics [11, 12]. Nonetheless, these perturbation method-based studies on nonlinear systems generate nonphysical results that are bound to be misinterpreted. The presentation that follows reviews the nonlinearity and nonstationary bifurcation of a softening Duffing oscillator from the time-frequency perspective established using IF. It is noted that, although IF is considered a viable tool for exploring nonlinear dynamic response, little effort has been made to study the generation and evolution of bifurcation to ultimate chaotic response, a process that is inherently nonstationary and transient. A Duffing oscillator and its linearized counterpart are studied first by fast Fourier transform (FFT), short time Fourier transform (STFT), Gabor transform, and instantaneous frequency (IF). The second part of the chapter presents an in-depth investigation into the route-to-chaos generated by the Duffing oscillator under nonstationary excitation using conventional nonlinear dynamic analysis tools and IF.

6.1 Implication of Linearization in Time-Frequency Domains

To examine the impact of linearization, the responses of a nonlinear Duffing oscillator and its linearized version under stationary excitation are investigated by FFT, time-frequency analysis tools, and IF. Analogous to complex nonlinear systems, including the rolling motion of a ship, Duffing oscillators have the advantage of simplicity and can be investigated in sufficient detail. Of interest is the response of a particular Duffing oscillator subject to a harmonic excitation with viscous damping, which has been found to exhibit hysteretic and chaotic behaviors [9]. The general form of the nondimensional Duffing oscillator is

$$\ddot{x} + 2\mu\dot{x} + \beta x + \alpha x^3 = a \cos(\omega t) \tag{6.1}$$

where μ, β, α, a, and ω are constants. When the motion is small, the cubic term can be linearized with respect to the equilibrium point zero and be ignored as

$$\ddot{x} + 2\mu\dot{x} + \beta x = a \cos(\omega t) \tag{6.2}$$

If $\beta - \mu^2 > 0$, the general solution can be simplified as

$$x(t) = \underbrace{A\, e^{-\mu t} \cos(\sqrt{\beta - \mu^2}\, t + \theta)}_{transient} + \underbrace{\frac{a}{\sqrt{(\omega^2 - \beta)^2 + 4\mu^2\omega^2}} \cos(\omega t - \theta)}_{steady-state} \tag{6.3}$$

When $\mu < 1$, it is an underdamped system and the damped natural frequency is $\sqrt{\beta - \mu^2}$. The frequency of the steady-state response ω is the same as the excitation frequency. In the following, the single-well Duffing oscillator investigated in [10] is adopted, with $\alpha = -1$, $\beta = 1$, $\mu = 0.2$, and the stationary excitation amplitude, a, being kept at 0.32 and the excitation frequency at 0.78 rad/s. Figure 6.1 shows the FFT of the linearized and nonlinear Duffing oscillators. It is hard to distinguish one from the other at first glance, and most would think both hold only one frequency at 0.12 Hz.

Two time-frequency analysis methods, short-time Fourier transform (STFT) and Gabor wavelet transform are applied to investigate nonlinear and linearized responses as follows. Neither the dominant frequency nor the nonlinear effect is precisely resolved by STFT in Figure 6.2. To improve the frequency resolution, one has to increase the width of the time window, thus inevitably resulting in poor time resolution. This dilemma is inherent to all Fourier-based time-frequency distributions. The Gabor wavelet transform [13] in Figure 6.3 shows a better time-frequency resolution, however it is still unable to differentiate the nonlinear response from the linearized one. As the only difference between these two spectra is the tiny irregular frequencies near the dominant frequency, which would be taken by most as the noise to be filtered, linearization would be adopted in a heartbeat. But the careless assumption that the response of the nonlinear system can be linearized, and afterwards controlled, without further investigation could expose the system to the potential risk of abrupt breakdown. Linearization, an approach generally accepted as the premise for dealing with nonlinear problems without caveat, in fact distorts the inherent underlying physical characteristics. A system could be falsely characterized, thus risking becoming unstable as a result.

IF provides an alternative look at the response in the simultaneous time-frequency domain. Figure 6.4(a) shows the selected time history (top) of the linearized Duffing oscillator along with its extracted IMF C1 (middle) and residual R(t) (bottom). IMF C1, the mode containing the highest frequency components, is characteristically similar to the original time response that is a harmonic oscillation. The response of the linearized Duffing oscillator has only one frequency, thus resulting in only one IMF. According to the study in [3], the IF is exactly the reciprocal of the period in the IMF mode. Hence the steady-state IF of the C1 mode indicates

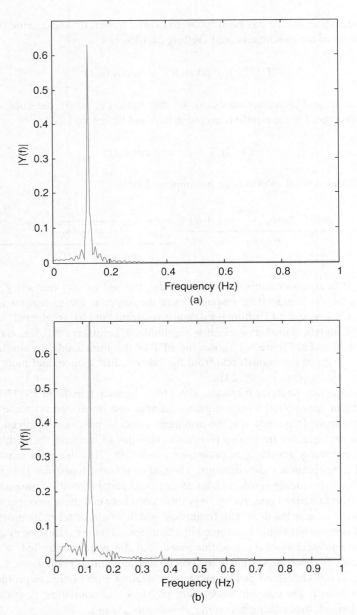

Figure 6.1 Fast Fourier transform of (a) linearized (b) nonlinear Duffing oscillator. Reproduced with permission from Meng-Kun Liu, C.Steve Suh, 2012, "Temporal and Spectral Responses of a Softening Duffing Oscillator Undergoing Route-to-Chaos", Nonlinear Science and Numerical and Numerical Simulation, (17) pp. 5217–5228. Copyright 2012 Elsevier

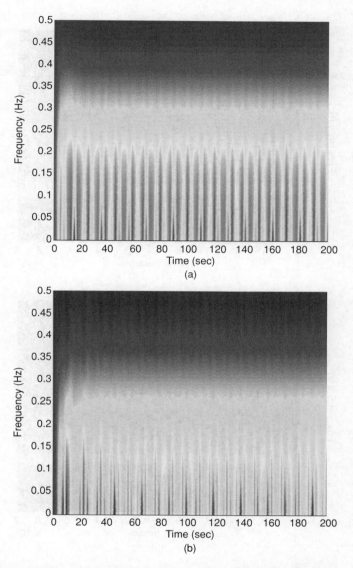

Figure 6.2   Short-time Fourier transforms of (a) linearized and (b) nonlinear Duffing oscillators under stationary excitation (Sampling frequency = 2 Hz). Reproduced with permission from Meng-Kun Liu, C.Steve Suh, 2012, "Temporal and Spectral Responses of a Softening Duffing Oscillator Undergoing Route-to-Chaos", Nonlinear Science and Numerical and Numerical Simulation, (17) pp. 5217–5228. Copyright 2012 Elsevier

a constant frequency at 0.124 Hz as seen in Figure 6.4(b), which coincides with the frequency $\omega = 0.78/2\pi$ Hz from the linearized model described in Equation (6.3).

The following analysis retains the cubic nonlinear term of the Duffing oscillator. Figure 6.5 gives the IMFs and IFs of the nonlinear Duffing oscillator in Equation (6.1). Although the IMF

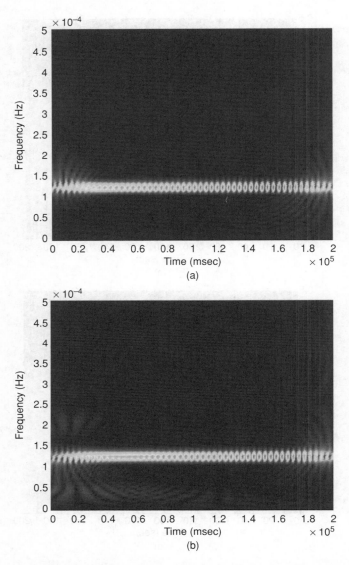

Figure 6.3 Gabor transforms of (a) linearized and (b) nonlinear Duffing oscillators under stationary excitation (Sampling frequency = 2 Hz). Reproduced with permission from Meng-Kun Liu, C.Steve Suh, 2012, "Temporal and Spectral Responses of a Softening Duffing Oscillator Undergoing Route-to-Chaos", Nonlinear Science and Numerical and Numerical Simulation, (17) pp. 5217–5228. Copyright 2012 Elsevier

C1 waveform in Figure 6.5(a) seems harmonic, the IF in Figure 6.5(b) shows that it is not, but rather displays a simultaneously temporal-modal behavior oscillating periodically between 0.11 and 0.14 Hz with the mean value at 0.124 Hz, which happens to be the frequency of the linearized system. The IF shows that the frequency of the nonlinear response is not static but is

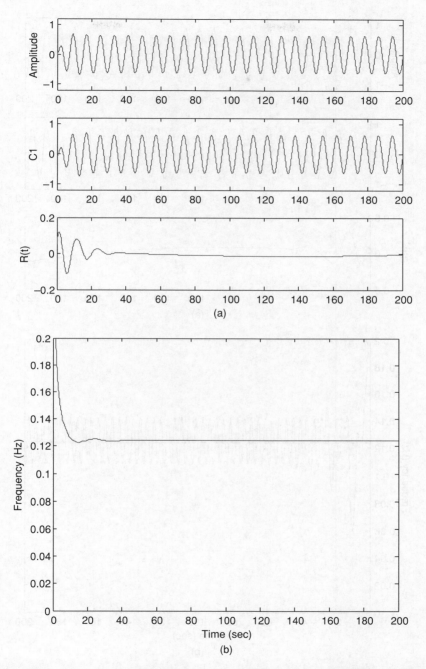

Figure 6.4 (a) Time response and its IMFs and (b) IF of the linearized Duffing oscillator. Reproduced with permission from Meng-Kun Liu, C.Steve Suh, 2012, "Temporal and Spectral Responses of a Softening Duffing Oscillator Undergoing Route-to-Chaos", Nonlinear Science and Numerical and Numerical Simulation, (17) pp. 5217–5228. Copyright 2012 Elsevier

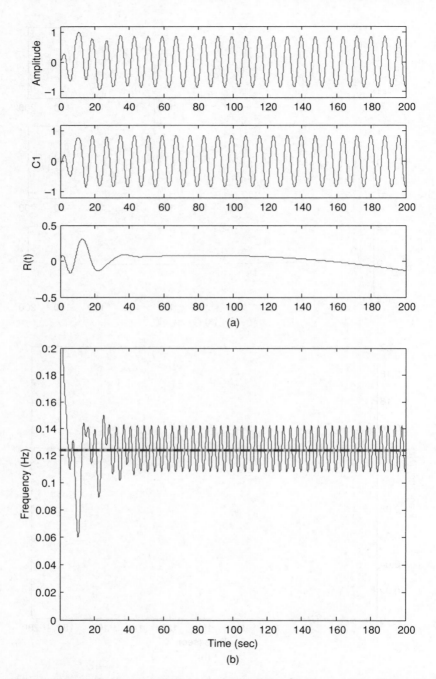

Figure 6.5 (a) Time response and its IMFs (b) IF of the nonlinear Duffing oscillator. Reproduced with permission from Meng-Kun Liu, C.Steve Suh, 2012, "Temporal and Spectral Responses of a Softening Duffing Oscillator Undergoing Route-to-Chaos", Nonlinear Science and Numerical and Numerical Simulation, (17) pp. 5217–5228. Copyright 2012 Elsevier

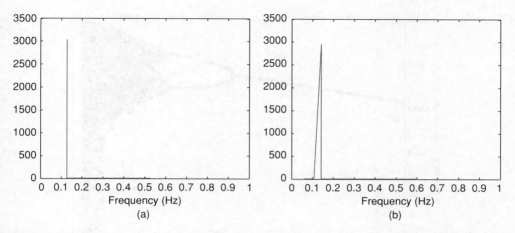

Figure 6.6 Marginal spectrum of (a) linearized and (b) nonlinear Duffing oscillator. Reproduced with permission from Meng-Kun Liu, C.Steve Suh, 2012, "Temporal and Spectral Responses of a Softening Duffing Oscillator Undergoing Route-to-Chaos", Nonlinear Science and Numerical and Numerical Simulation, (17) pp. 5217–5228. Copyright 2012 Elsevier

rather varying within a certain range. This is further asserted by reviewing the marginal spectrum, which shows frequency distribution in the probabilistic sense. The marginal spectrum in Figure 6.6(a) shows that there is a single frequency associated with the linearized Duffing oscillator. But for the nonlinear case in Figure 6.6(b) there are multiple frequencies between 0.11 and 0.14 Hz. Since a marginal spectrum is the occurrence probability of frequency components over a selected time span, it is evident that there are multiple frequency components in the response.

Two observations can be made. First, the Fourier spectrum is unable to reveal the true characteristic of the nonlinear response. Second, by comparing the marginal spectra of the nonlinear and linearized responses, it is observed that linearization misinterprets nonlinear features, replacing multiple frequencies with a single frequency. The false representation of spectral characteristics implies that the common frequency-domain-based controllers designed using linearization would misinterpret the frequency response, thus being incapable of realizing the ongoing evolution of bifurcation. Since route-to-chaos is a transient progress in which spectral response deteriorates from being periodic to aperiodic and broadband, linearization and Fourier-based controller design would most certainly fail to identify the inception of bifurcation and chaos, and the stability bound of the system.

6.2 Route-to-Chaos in Time-Frequency Domain

Unlike the nonlinear response due to stationary excitation, the one induced by nonstationary excitation is a temporal transition from bifurcation to chaos with a time-varying spectrum. Conventional tools including bifurcation diagram, phase portrait and Poincaré map are used to investigate the route-to-chaos in contrast to the result obtained by IF. The Duffing oscillator of single potential well in Equation (6.1) is again considered, but with a time-increasing excitation amplitude. When the amplitude is small, the response is bounded and remains in the valley of the potential well. When the external excitation gradually increases, the response, whose amplitude may still be bounded, could jump to an unbounded solution after it passes a critical limit – a phenomenon similar to the catastrophic capsizing of a marine vessel. The excitation

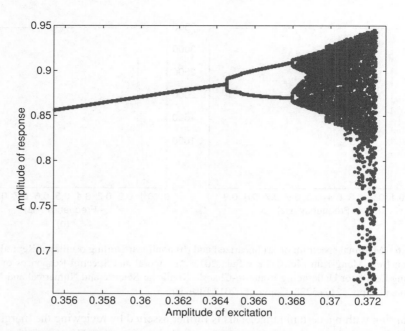

Figure 6.7 Nonstationary bifurcation diagram with increasing excitation amplitude. Reproduced with permission from Meng-Kun Liu, C.Steve Suh, 2012, "Temporal and Spectral Responses of a Softening Duffing Oscillator Undergoing Route-to-Chaos", Nonlinear Science and Numerical and Numerical Simulation, (17) pp. 5217–5228. Copyright 2012 Elsevier

amplitude in Equation (6.1) is considered as the control parameter, with all other coefficients ($\alpha = -1$, $\beta = 1$, and $\mu = 0.2$) following from [12], where a stationary bifurcation diagram of a Duffing oscillator was generated using multiple scales and Floquet theory. Again the excitation frequency is held at 0.78 rad/s. The excitation amplitude is a linear time function, $a = 0.32 + 10^{-7}t$. Thus the response is bounded within the potential well, and the slow increase of the amplitude ensures that the bifurcation process can be clearly observed. By making the amplitude a function of time, a nonstationary bifurcation diagram is constructed in Figure 6.7. While agreeing with the result in [7], the figure also indicates the penetration effect and the elimination of the stationary discontinuities.

In general the bifurcation depicted in Figure 6.7 can be divided into three stages before becoming unbounded. The stage of singular frequency is the first stage that existed when the excitation was first applied. At this stage each value of the excitation amplitude refers to only one point in the bifurcation diagram, meaning that there is only one frequency at any time instance, though this frequency may not be static. When the excitation amplitude increases, period-doubling is observed. When multiple periods show up, it is the third, fractal stage. Within this stage spectral bandwidth increases, but remains bounded. When excitation amplitude exceeds 0.3725, the response becomes unbounded. In addition to the bifurcation diagram, phase portraits and Poincaré maps are plotted to help resolve the bifurcation-to-chaos progression in time. Figure 6.8(a) shows a close-trajectory phase portrait and a corresponding single-dot Poincaré map, representing a periodic response with single frequency. A 2T period-doubling bifurcation and a 4T period-doubling bifurcation are seen in Figures 6.8(b) and (c), respectively. Fractal structures emerge from Figure 6.8(d) and deteriorate in Figure 6.8(e)

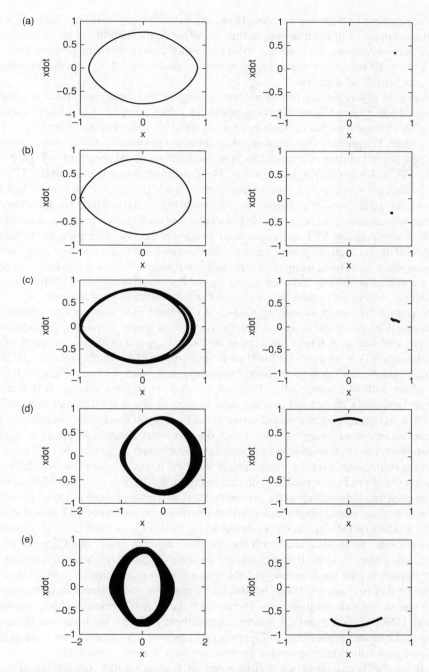

Figure 6.8 Phase portrait (left) and Poincaré map (right) for (a) periodic motion (b) 2T period-doubling bifurcation (c) 4T period-doubling bifurcation (d) fractal (e) fractal before being unbounded. Reproduced with permission from Meng-Kun Liu, C.Steve Suh, 2012, "Temporal and Spectral Responses of a Softening Duffing Oscillator Undergoing Route-to-Chaos", Nonlinear Science and Numerical and Numerical Simulation, (17) pp. 5217–5228. Copyright 2012 Elsevier

before the response becomes unbounded. However, it is difficult to make out the motion states due to the overlapping of the trajectories, thus highlighting the inability of phase portraits and Poincaré maps to capture the transient phenomena of the route-to-chaos process. Other than showing the qualitative transition from bifurcation to chaos, the bifurcation diagram does not provide any further information.

Responses in Figure 6.8 are further evaluated using FFT and marginal spectrum. The FFT of the periodic motion in Figure 6.9(a) indicates a frequency component at 0.142 Hz, while the marginal spectrum shows that the frequency is not static but rather oscillates between 0.11 and 0.14 Hz, which is a primary characteristic of a nonlinear response. As the excitation amplitude increases, a second cluster of frequencies appears in the marginal spectrum in Figure 6.9(b), while the FFT still shows a single frequency. The same observation is made of the 4T period-doubling bifurcation in Figure 6.9(c). Furthermore, the marginal spectra in Figures 6.9(d) and (e) indicate a high probability of frequency occurrence between 0 and 0.05 Hz. It indicates that the response frequency proliferates to be broadband and undergoes a route-to-chaos process, while the corresponding FFT is incapable of resolving the changing process. In addition, noticeable fictitious high frequencies emerge in Figures 6.9(b), (d), and (e). These artificial high frequencies stem from using superharmonic components to fit the dynamic response in FFT – a result of averaging and eliminating the subharmonic frequencies of the real signal. The result shows that FFT could be nonphysical and misinterpret the true response.

Instantaneous frequency is then applied to the selected time segments to scrutinize the three stages that are characterized, respectively, by a singular frequency, period-doubling bifurcation, and fractal. A fixed time window of 200 s is applied to ensure a better resolution. Several figures follow to illustrate the time-progression of the periodic motion, 2T period-doubling bifurcation, 4T period-doubling bifurcation, and route-to-chaos. Figure 6.10 shows the IFs along with its corresponding IMFs of the singular frequency stage. It is found that this stage, previously considered as a periodic motion by phase portrait and Poincaré map, actually has oscillating frequencies indicative of nonlinearity. It is surprising to compare the IF of the one under nonstationary excitation with the one under stationary excitation. Although the nonstationary excitation amplitude is increased slowly with a rate of only 10^{-7} per second (whose time responses/waveforms are similar to those in Figure 6.5), their IFs show differences. Comparing the IFs in Figure 6.10(b) with that in Figure 6.5(b), it is clear that IF is capable of resolving even the slightest shift of the system that was not revealed using all previous methods.

As the excitation amplitude increases further, the response undergoes a 2T period-doubling bifurcation. The first IMF mode, C1, is similar to the time response, and the corresponding IF looks identical to the IF associated with the singular frequency stage. The C2 mode emerges with increasing amplitude. Its IF in Figure 6.11 indicates a second frequency oscillating about 0.06 Hz, which is half the frequency of the first mode, thus a period-doubling bifurcation. However, neither frequency is static, but displays a temporal-modal structure oscillating periodically and shows rich nonlinearities. Figure 6.12 shows a 4T period-doubling bifurcation. Two more IMF modes, C3 and C4, emerge. All of them look like harmonic oscillations, but none possess constant periodicity. In Figure 6.12(b), two new frequencies are generated with a less vigorous oscillation compared to the first two instantaneous frequencies.

The IF of the fractal structure is illustrated in Figure 6.13(b). Unlike the IFs in the 4T period-doubling bifurcation, where relatively regular temporal-modal oscillations were observed, no IFs follow a well-defined pattern or structure. All IMF modes, C1, C2, C3,

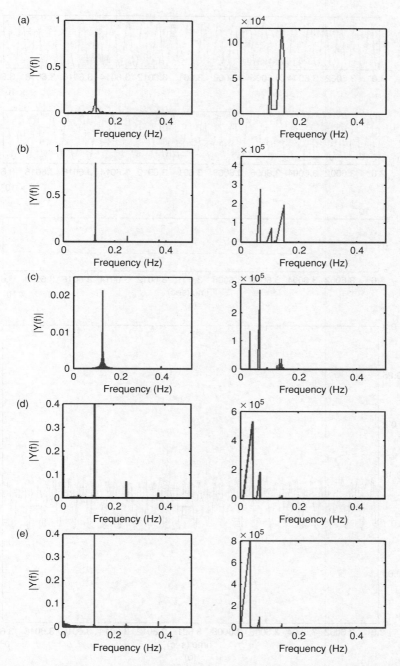

Figure 6.9 FFT (left) and marginal spectrum (right) for (a) periodic motion (b) 2T period-doubling bifurcation (c) 4T period-doubling bifurcation (d) fractal (e) fractal before being unbounded. Reproduced with permission from Meng-Kun Liu, C.Steve Suh, 2012, "Temporal and Spectral Responses of a Softening Duffing Oscillator Undergoing Route-to-Chaos", Nonlinear Science and Numerical and Numerical Simulation, (17) pp. 5217–5228. Copyright 2012 Elsevier

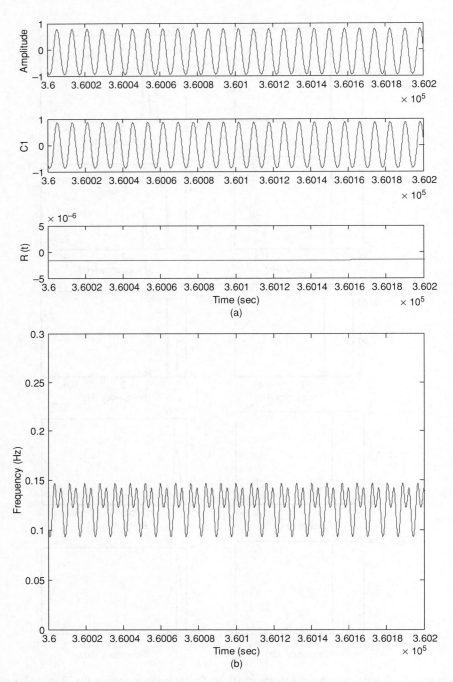

Figure 6.10 (a) Time response and its IMFs (b) Instantaneous frequencies of the stage of singular frequency (single frequency for each time point). Reproduced with permission from Meng-Kun Liu, C.Steve Suh, 2012, "Temporal and Spectral Responses of a Softening Duffing Oscillator Undergoing Route-to-Chaos", Nonlinear Science and Numerical and Numerical Simulation, (17) pp. 5217–5228. Copyright 2012 Elsevier

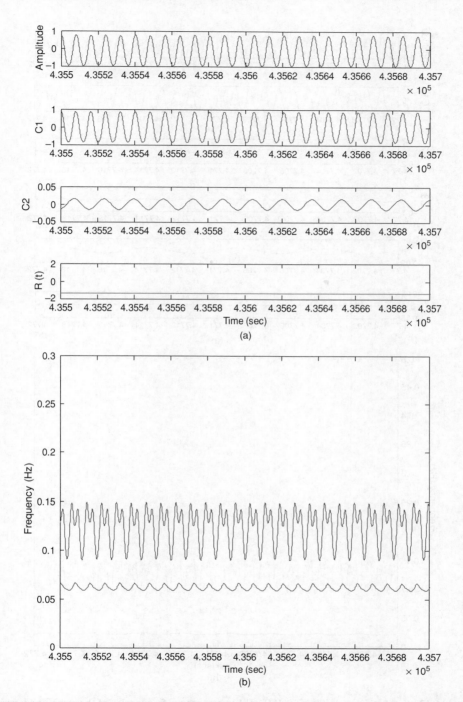

Figure 6.11   (a) Time response and its IMFs (b) Instantaneous frequencies of the 2T period-doubling bifurcation. Reproduced with permission from Meng-Kun Liu, C.Steve Suh, 2012, "Temporal and Spectral Responses of a Softening Duffing Oscillator Undergoing Route-to-Chaos", Nonlinear Science and Numerical and Numerical Simulation, (17) pp. 5217–5228. Copyright 2012 Elsevier

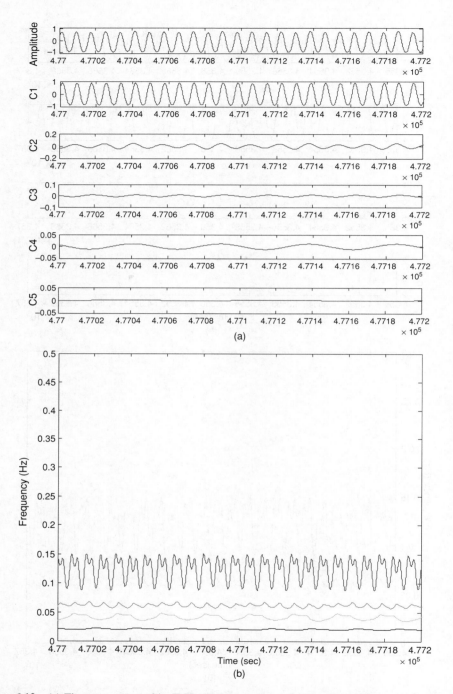

Figure 6.12 (a) Time response and its IMFs (b) Instantaneous frequencies of the stage of 4T period-doubling bifurcation. Reproduced with permission from Meng-Kun Liu, C.Steve Suh, 2012, "Temporal and Spectral Responses of a Softening Duffing Oscillator Undergoing Route-to-Chaos", Nonlinear Science and Numerical and Numerical Simulation, (17) pp. 5217–5228. Copyright 2012 Elsevier

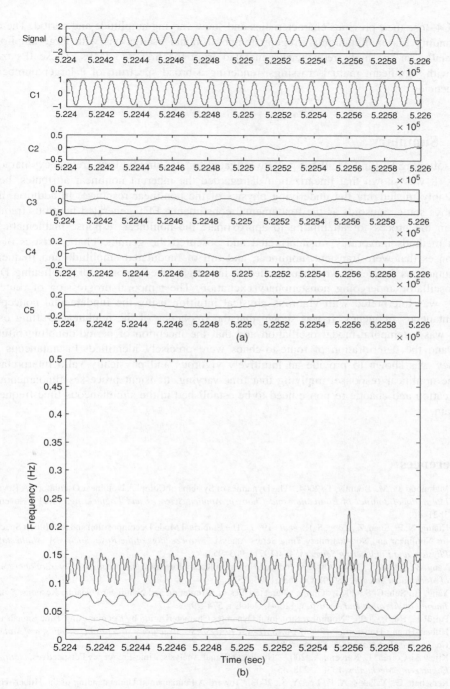

Figure 6.13 (a) Time response and its IMFs (b) Instantaneous frequencies of the stage of 4T period-doubling bifurcation. Reproduced with permission from Meng-Kun Liu, C.Steve Suh, 2012, "Temporal and Spectral Responses of a Softening Duffing Oscillator Undergoing Route-to-Chaos", Nonlinear Science and Numerical and Numerical Simulation, (17) pp. 5217–5228. Copyright 2012 Elsevier

and C4, lose their previous characteristics and show varying amplitude and period. The corresponding IFs exhibit highly irregular temporal-modal oscillations with large amplitude, especially for the high frequencies generated from the first two IMFs. These IFs oscillate with significant mutual-crossings, rendering a broad spectrum of a larger number of frequencies.

6.3   Summary

The softening Duffing oscillator was investigated for its intrinsic time-frequency characteristics. It was shown that linearization disregarded the inherent nonlinear attributes, hence inevitably misjudging the underlying physics of the nonlinear response. In addition, as it employs static sinusoids as the fundamental constituents, FFT generated fictitious frequencies in the process of attempting to approximate the nonlinear response mathematically. Both methods inexorably misinterpreted and obscured the genuine characteristics of the responses that were transient, nonlinear, and full of modulated amplitude and frequency. Instantaneous frequency was then applied to investigate the responses of a softening Duffing oscillator undergoing nonstationary excitation. The temporal progression of route-to-chaos was interpreted with vigorous physical intuition using the fundamental concept of instantaneous frequency. It was shown that the frequency of the nonlinear Duffing oscillator was a temporal-modal oscillation and that the inception of period-doubling bifurcation and the deterioration of route-to-chaos were precisely identified. Instantaneous frequency was shown to provide an intuitively vigorous and physically valid interpretation of the nonlinear response, implying that time-varying, transient processes fundamental of bifurcation and chaotic response need to be established in the simultaneous time-frequency domain.

References

[1] Mahmoud, G. M., Bountis, T., 2004, "The Dynamics of Systems of Complex Nonlinear Oscillators: A Review," *International Journal of Bifurcation and Chaos in Applied Sciences and Engineering*, 14(11), November, 3821–46.
[2] Huang, N. E., Shen, Z., Long, S. R., *et al.*, 1998, "The Empirical Mode Decomposition and the Hilbert Spectrum for Nonlinear and Non-stationary Time Series Analysis," *Proceedings of the Royal Society A: Mathematical, Physical and Engineering Sciences*, 454(1971), 903–95.
[3] Yang, B., Suh, C. S., 2006, "On Fault Induced Nonlinear Rotary Response and Instability," *International Journal of Mechanical Science*, 48(10), October, 1103–25.
[4] Yang, B., Suh, C. S., 2004, "On the Characteristics of Bifurcation and Nonlinear Dynamic Response," *ASME Journal of Vibration and Acoustics*, 126(4), October, 574–79.
[5] Pai, P. F., Hu, J., 2006, "Nonlinear Structural Dynamics Characterization by Decomposing Time Signals using Hilbert-Huang Transform," *47th AIAA/ASME/ASCE/AHS/ASC Structures, Structural Dynamics, and Materials Conference, 1–4 May 2006, Newport, RI.*
[6] Kijewski-Correa, T., Kareem, A., 2007, "Nonlinear Signal Analysis: Time-Frequency Perspectives," *Journal of Engineering Mechanics*, 133(2), 238–45.
[7] Kerschen, G., Vakakis, A. F., Lee, Y. S., 2008, "Toward A Fundamental Understanding of the Hilbert-Huang Transform in Nonlinear Structural Dynamics," *Journal of Vibration and Control*, 14(1–2), 77–105.
[8] Douka, E., Hadjileoniadis, L.J., 2005, "Time-Frequency Analysis of the Free Vibration Response of A Beam with A Breathing Crack," *NDT&E International*, 38(1), 3–10.

[9] Pai, P. F., Palazotto, A. N., 2008, "Detection and Identification of Nonlinearities by Amplitude and Frequency," *Mechanical Systems and Signal Processing*, 22(5), 1107–32.

[10] Nayfeh, A. H., Sanchez, N. E., 1989, "Bifurcations in A Forced Softening Duffing Oscillator," *International Journal of Nonlinear Mechanics*, 24(6), 483–97.

[11] Moslehy, F. A., Evan-Iwanowski, R. M., 1991, "The Effects of Non-Stationary Processes on Chaotic and Regular Responses of The Duffing Equation," *International Journal of Nonlinear Mechanics*, 26(1), 61–71.

[12] Lu, C. H., Evan-Iwanowski, R. M., 1994, "Period Doubling Bifurcation Problems in The Softening Duffing Oscillator with Nonstationary Excitation," *Nonlinear Dynamics*, 5(4), 401–20.

[13] Gabor, D., 1946. "Theory of Communications," *Journal of the Institute of Electrical Engineers*, 93(26), 429–57.

7

Simultaneous Time-Frequency Control of Dynamic Instability

While most chaos control theories focus on controlling "static chaos," the deterioration of cutting stability via route-to-chaos is in fact "dynamic" – a transient, nonstationary process. A chaotic response is naturally bounded in the time-domain while in the meantime becoming unstably broadband in the frequency-domain. In the present chapter several control theories, either designed in the time-domain or frequency-domain, are reviewed. They are insufficient in addressing the route-to-chaos process. *A priori* knowledge of the system must be available for the control theories to work properly. They all demonstrate failure when the system state undergoes severe changes. It is concluded in the first half of the chapter that it is necessary for a viable controller to be adaptive and able to identify the system and facilitate proper control in real-time. Two control theories, the OGY method and Lyapunov-based controller, are applied to the stationary/nonstationary Hénon map and Duffing equation, respectively. They are successful in controlling autonomous, stationary systems but fail when the systems are nonautonomous and nonstationary. Based on the fundamental knowledge established in the present and previous chapters, a novel chaos control scheme is developed, having features that address the fundamental characteristics common to chaotic systems. Multiresolution analysis realized by filter banks that decompose a signal into its high frequency and low frequency components is incorporated. Built in the wavelet domain, the controller renders simultaneous manipulation in both the time and frequency domains. On-line identification and feedforward control are implemented via a revised version of the FXLMS algorithm discussed in Chapter 3. The control methodology developed in the present chapter is able to mitigate dynamical deterioration in both the time and frequency domains and properly regulate the response with the desired reference signal.

7.1 Property of Route-to-Chaos

Research on chaos control has drawn much attention over several decades. Open-loop control and closed-loop control are the two major categories. Open-loop control, which alters the behavior of a nonlinear system by applying a properly chosen input function or external excitation, is simple and requires no sensors. However, open-loop control is in general limited by the fact that its action is not goal oriented [1]. Closed-loop control, on the other hand, feedbacks a perturbation selected based upon the state of the system to control a prescribed dynamics. Of the many closed-loop chaos control theories formulated over the years, the OGY method, delayed feedback control, Lyapunov-based control, and adaptive control are considered prominent. The OGY method [2] uses small discontinuous parameter perturbation to stabilize a chaotic orbit and forces the trajectory to follow a target UPO (unstable periodic orbit) in a chaotic attractor. It uses the eigenvalues of the system's Jacobian at fixed point(s) to establish stability. But for chaotic systems of higher dimensions, there are complex eigenvalues or multiple unstable eigenvalues, making it difficult to control such systems by the OGY method [3]. Several revisions have been made to control chaos in higher-order dynamical systems [3, 4, 5, 6]. Another disadvantage is that the available adjustable range of the controlling parameter is limited by the distance between the system state variable and UPO. Because the initiation of OGY control requires that the state variable approaches the proximity of the target UPO, the waiting time can be shortened by applying the reconstruction of phase plane [7]. Nonetheless, it is very difficult to obtain an exact, analytic formula for an UPO. It is even more difficult to physically implement UPOs due to the unstable nature of such orbits [8]. Since the corrections of the parameter are discrete, rare, and small, presence of noise can lead to occasional bursts of the system into regions far from the desired periodic orbit [9]. These difficulties limit the OGY method to only a few applications such as the control of robot arms [10], forced pendulum [11, 12], and power systems [13].

Another widely accepted chaos control theory is the delayed feedback control (DFC) [9]. The stabilization of UPO of a chaotic system can be achieved either by combined feedback with the use of a specially designed external oscillator, or by delayed self-controlling feedback. The feedback is a small continuous perturbation that is less vulnerable to noise. Unlike the OGY method, it doesn't need *a priori* analytical knowledge of the system dynamic, except for the period of the target UPO, and it can be applied to a high-dimensional system. Recent efforts include stability analysis [14] and the stabilization of UPO with an arbitrarily large period [15]. A comprehensive review of the delayed feedback control method is found in [16]. The drawback of the delayed feedback control is that it is hard to conduct linear stability analysis of the delayed feedback system [16] and that its performance is very sensitive to the choice of the delay [17]. If the control goal is to stabilize a forced T-periodic solution, the delay will mandatorily be set to T. Alternatively, a heuristic method is used to estimate the delay time, but it is still difficult to find the smallest period.

Lyapunov's direct method allows the stability of a system to be determined without explicitly integrating the dynamic equation. It relies on the physical property that the system, whose total energy is continuously being dissipated, must eventually end up at an equilibrium point [18]. Suppose a continuously differentiable positive definite (Lyapunov) function $V(x)$ can be derived from the system. The system is stable if the derivative of $V(x)$ is negative semi-definite, and asymptotically stable if the derivative of $V(x)$ is negative definite. This concept can be integrated into controller design. As long as the derivative of a Lyapunov function of the system is confined to negative semi-definite or definite along the closed-loop system trajectory, the system is guaranteed stable or asymptotically stable. Lyapunov-based controller has been applied to the synchronization of chaos [19, 20, 21], Duffing oscillators [22, 23], chaotic pendulum [24] and robotics [18]. Its drawback is that the Lyapunov function cannot necessarily be asserted from some particular models, and the chosen parameters may be too conservative, thereby compromising the transient response of the system [18].

An identification algorithm usually is coupled with the control algorithm to facilitate adaptive control over the dynamical system that has unknown parameters in its governing equation. In adaptive control, parameter estimation and control are performed simultaneously. When the system parameters are estimated and control action is calculated based on the estimated parameters, the adaptive control scheme is called indirect adaptive control. In direct adaptive control, on the other hand, controller parameters are directly updated using an adaptive law. Adaptive control theory modifies the control law to cope with the time-varying parameters of the system. Even though adaptive control describes a nonlinear system by a linear model, the feedback tuning of its parameters renders the overall system response nonlinear. Hence, adaptive control is widely adopted for the control of chaotic systems, such as Hénon map [25], nonlinear pendulum [26], chaos synchronization [27], and hyper-chaos system [28]. *A priori* knowledge of the system is required for model-based adaptive control, which focuses exclusively on time-domain performance.

Although all demonstrated capability in controlling chaos, however, the applicability of the chaos control theories reviewed above is limited to autonomous, stationary systems. They are all developed assuming chaotic system to be autonomous even though nonlinear dynamics concern predominantly with nonautonomous systems. These controllers are good at handling "static chaos" meaning the state of chaos doesn't change. "Static chaos" doesn't transition from bifurcation to chaos as is explicit in a bifurcation diagram. The spectral bandwidth of its response doesn't change either. For nonautonomous, nonstationary systems whose chaotic responses are "dynamic," these chaos control theories would fail. Two examples on chaos control are studied in the followings.

7.1.1 OGY Control of Stationary and Nonstationary Hénon Map

Consider the two-dimensional iterative (Hénon) map function below with $\alpha > 0$, $|\beta| < 1$, and $r_n^2 = x_n^2 + y_n^2$

$$\begin{cases} x_{n+1} = 1 + y_n - \alpha x_n^2 \\ y_{n+1} = \beta x_n \end{cases} \qquad (7.1)$$

When parameter α is increasing, the system undergoes bifurcation, as depicted in Figure 7.1. Because α varies in time, the nonlinear dynamics shown in Figure 7.1 is nonautonomous and time-variant. If $\beta = 0.4$ and α is set at 1.2, the response is a chaotic attractor. But this time it is a case of "static chaos." The range of its time response and frequency spectrum remains unchanging. Figure 7.2 shows the corresponding time response and instantaneous frequency (IF) [29] of the Hénon map when the OGY method [30] is applied. With the controller being turned on at the 300th-time step in Figure 7.2(a), the chaotic response is stabilized to a fixed point. The response is examined by IF in Figure 7.2(b) using an integration time step $t = 0.1$ sec. The IF plot shows a transient between $t = 30$ sec and $t = 62$ sec, followed by a null region devoid of any time-frequency activity. Both the time response and IF signify that the OGY method is able to stabilize the chaotic attractor generated by a stationary Hénon Map.

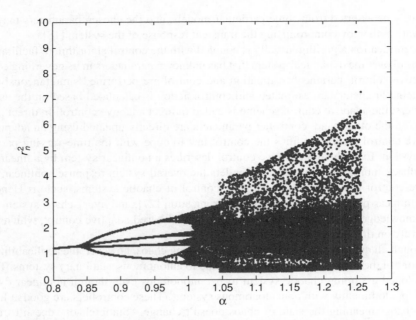

Figure 7.1   Bifurcation diagram of Hénon map. Reproduced with permission from Meng-Kun Liu, C. Steve Suh, 2012, "Simultaneous Time-Frequency Control of Bifurcation and Chaos," Communications in Nonlinear Science and Numerical and Numerical Simulation, (17) pp. 2539–2550. Copyright 2012 Elsevier

Figure 7.3 shows the bifurcation diagram of a nonstationary Hénon Map controlled by the same OGY method, in which the controlling parameter α is increasing in time. The OGY method fails for such a nonautonomous, nonstationary system. In the figure, the controller is activated when $\alpha = 1.2$. For a stationary system, r_n^2 would be fixed when the controller is turned on, which means that it has only one frequency when the system is under control. But in Figure 7.3 it shows that the trajectory in the bifurcation diagram is no longer a horizontal line. It means that the frequency is changing and that chaos abruptly emerges when α is increased to around 2.3. The result is not surprising because the concept of the OGY method is based on the linearization of the Poincaré map. Hence its stability region is inevitably limited to the vicinity of the equilibrium point.

7.1.2   Lyapunov-based Control of Stationary and Nonstationary Duffing Oscillator

Consider the following two Duffing oscillators

$$\ddot{x} + p\dot{x} + p_1 x + x^3 = q \cos(\omega t) \tag{7.2}$$

$$\ddot{x} + p\dot{x} + p_1 x + x^3 = (q + 0.01t) \cos(\omega t) \tag{7.3}$$

with $p = 0.4, p_1 = -1.1, q = 2.1$, and $\omega = 1.8$. Note that the excitation amplitude is a time function in Equation (7.3). The Lyapunov-based controller designed in [23] is applied to both cases

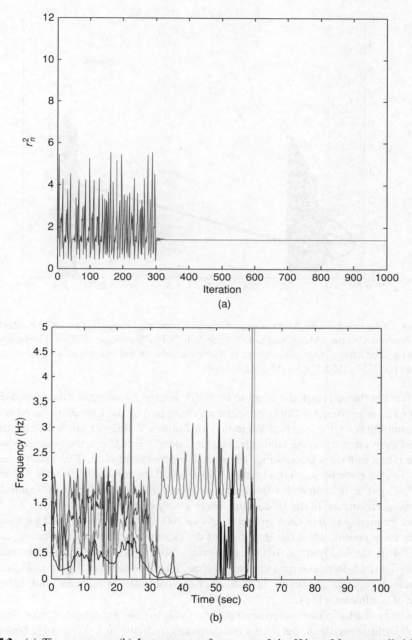

Figure 7.2 (a) Time response (b) Instantaneous frequency of the Hénon Map controlled by OGY method. Reproduced with permission from Meng-Kun Liu, C. Steve Suh, 2012, "Simultaneous Time-Frequency Control of Bifurcation and Chaos," Communications in Nonlinear Science and Numerical and Numerical Simulation, (17) pp. 2539–2550. Copyright 2012 Elsevier

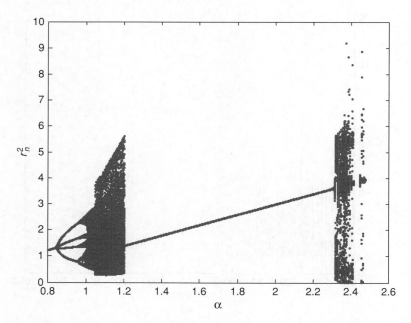

Figure 7.3   Bifurcation diagram of a nonautonomous Hénon map controlled by OGY method. Reproduced with permission from Meng-Kun Liu, C. Steve Suh, 2012, "Simultaneous Time-Frequency Control of Bifurcation and Chaos," Communications in Nonlinear Science and Numerical and Numerical Simulation, (17) pp. 2539–2550. Copyright 2012 Elsevier

and a reference (target) response is set to be $sin(t)$. Figure 7.4 compares the responses of the stationary and nonstationary Duffing oscillators controlled by the Lyapunov-based controller.

The amplitudes of the external excitation in Figures 7.4(a)–(c) are held constant. Figure 7.4(a) shows that when the controller is turned on at $t = 500$ sec, the controller stabilizes the system and mitigates the chaotic response to a periodic motion. This result agrees with its time-domain error between the system response and the reference trajectory as shown in Figure 7.4(b). The instantaneous frequency in Figure 7.4(c) shows that the controller also has a good performance in the IF domain. There's only one frequency left after $t = 500$ sec. A second frequency is also seen emerging at $t = 700$ sec. Conversely, the Lyapunov-based controller loses control when the amplitude of the external excitation is increasing in time in Figures 7.4(d)–(f). In Figure 7.4(d) the response is no longer a periodic motion and in Figure 7.4(e) the time-domain error increases in time. Further, from the IF plot in Figure 7.4(f), the frequency remains oscillating in time after the controller is engaged, indicative of the presence of nonlinearity [31].

In summary, the Lyapunov-based controller was able to deny the progression of chaos in the stationary Duffing oscillator; but for the nonstationary oscillator, it lost control in both the time and frequency domains. The review performed above can be summarized by the following observations:

1. All reviewed chaos control theories failed to mitigate "dynamic chaos." The OGY method is based on the linearization of the Poincaré map. It discards nonlinear terms and uses a Jacobian matrix to determine the stability of the equilibrium point around its vicinity. The

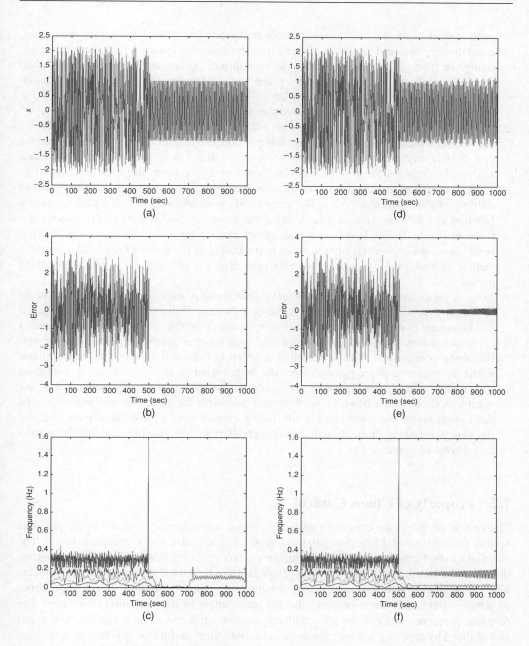

Figure 7.4 (a) Time response, (b) time-domain error, and (c) IF of a stationary Duffing oscillator, and (d) time response, (e) time-domain error, and (f) IF of a nonstationary Duffing oscillator, all controlled by a Lyapunov-based controller. Reproduced with permission from Meng-Kun Liu, C. Steve Suh, 2012, "Simultaneous Time-Frequency Control of Bifurcation and Chaos," Communications in Nonlinear Science and Numerical and Numerical Simulation, (17) pp. 2539–2550. Copyright 2012 Elsevier

method worked for the autonomous, stationary system but failed for the nonautonomous, nonstationary system because the latter's trajectories in the Poincaré map were subject to change in time. They ran away from the equilibrium point and the stability established by the Jacobian matrix at equilibrium point was no longer valid. The Lyapunov-based controller also failed for the nonautonomous, nonstationary system because the variation of the system parameter directly affected the derivation of the Lyapunov function.

2. *A priori* knowledge of the system must be available for the control theories to be effective. For the OGY method, the period of unstable periodic orbits (UPOs) must be known and the system state on the Poincaré map must be observed. But it is very difficult to obtain exact analytic formulation for a UPO [8]. The Lyapunov-based controller uses an energy-like concept to define the Lyapunov function, and examines its derivative, to determine the stability of the equilibrium point. But there is no systematic way of finding the Lyapunov function and in some cases it's basically a matter of trial and error. Even if a Lyapunov function can be found, it is only for an autonomous, nonstationary system. For delayed feedback control, the delayed time is set to the period of the desired orbit, and a heuristic method is then used to estimate the delay time. But it is still difficult to find the smallest period.

3. None of the theories control both time and frequency responses simultaneously. Except for the OGY method, all are formulated in the time domain. However, as seen previously, the instantaneous frequency of a nonlinear, nonstationary system undergoing route-to-chaos is characteristically time-modulated and has broad spectral bandwidth with emerging new frequency components. This was further asserted in the nonstationary Duffing oscillator where both time and frequency responses deteriorated at the same time. A controller designed to control time-domain error would not be able to negate the increasing of the spectrum. On the other hand, a controller designed in the frequency domain would confine the expansion of the bandwidth while losing control over time-domain error. Neither frequency-domain nor time-domain-based controllers are effective in mitigating bifurcation and chaotic response.

7.2 Property of Chaos Control

The review on the chaos control of nonautonomous, nonstationary systems in the previous section provides several hints essential to the development of a viable control solution. The solution can be formulated by recognizing the various attributes inherent in a chaotic system, including the simultaneous deterioration of dynamics in both the time and frequency domains when it bifurcates, non-stationarity, and sensitivity to initial conditions. For a linear time-invariant system, only the amplitude and the phase angle of the excitation input vary. The response frequency remains the same with respect to the input frequency, and the system can be stabilized by applying a proper feedback gain. Both time- and frequency-domain responses are bounded. However, this is not the case for the chaotic response generated by a chaotic system, which contains an infinite number of unstable periodic orbits of all periods called strange attractors. Chaotic response doesn't remain following one periodic orbit but switches rapidly between many unstable periodic orbits. If the chaotic response is projected into a Poincaré section, a lower dimensional subspace transversal to the trajectory of the response, it can be shown that the intersection points congregate densely and are confined within a finite area. It indicates that the chaotic response is bounded in the time domain while simultaneously becoming unstably broadband in the frequency domain due to the rapid switching between infinite numbers of UPOs. Hence, for a chaos control algorithm to be effective, control has to be performed in the time- and frequency-domains simultaneously.

The second property universal of chaotic systems is nonstationarity. Route-to-chaos is a temporal, transient process. The location and the stability of the equilibrium point therefore also vary in time. For a high-dimensional system, *a priori* knowledge of the system is often hard to come by. It is thus necessary for a viable chaos control scheme to conduct online identification and control at the same time in order to cope with the time-varying parameters of the system. The third property is the sensitivity of a chaotic system to initial conditions. A minor deviation between two closed initial trajectories might diverge exponentially with the increase of time, thus implying that a small perturbation could render the system unstable. Reversely, a nonlinear system can also be stabilized by a small perturbation, as implied by the open-loop chaos control theories of the early days in which chaotic systems were stabilized by giving small perturbations to their input or system parameters.

A solution with physical features effective in addressing the identified properties is described below. To address the need for providing simultaneous time and frequency resolution, Parseval's theorem is turned to for inspiration, which states that the total energy computed in the time-domain equals the total energy computed in the frequency-domain, thus implying that it's possible to incorporate both time-domain control and frequency-domain control. Wavelet transform (WT) localizes a time event and detects the ensuing changes in the wavelet domain, which is essentially a simultaneous time-frequency domain. Unlike Fourier analysis that approximates signals using sinusoids, WT uses finite, compact-supported orthogonal functions and provides localized time and frequency resolution through the translation and dilation of a base wavelet function. Efforts have been reported on incorporating discrete WT in control theory that explored (wavelet) multi-resolution through employing iterative filter banks. A wavelet filter bank runs a signal through two parallel channels, filtering one channel with a high-pass filter and the other with a low-pass filter, and then down-samples the signal by two. Control algorithms presented in Refs. [32, 33, 34, 35] were all constructed using wavelet coefficients by multiresolution analysis. They all reported improved transient performance. Discrete wavelet transform (DWT) was incorporated with a neural network to indentify unknown systems in real-time [36, 37, 38, 39, 40, 41], where adaptive control rules were applied. Relieved computational load and higher accuracy for system identification were among the benefits. Reference [42] applied DWT to replace the longstanding higher-order Taylor-series approach. Such a method reduces complexity and increases efficiency, but it is only applicable for linear models. In general, the coefficients of multiresolution analysis inherently carry simultaneous time-frequency information. Additionally, down-sampling in filter bank operation greatly reduces the amount of data and thus shortens the computation time.

7.2.1 Simultaneous Time-Frequency Control

To address the nonstationary nature of a chaotic system the concept of active noise control is adopted. Active noise control puts a control algorithm driven loudspeaker near the sound source to attenuate the sound. The sound source is cancelled by the sound emitted by the loudspeaker, which has the same amplitude as the source but of an opposite phase. The most

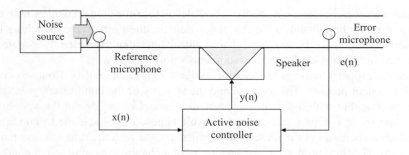

Figure 7.5 Signal channel broadband feedforward active noise control system in a duct [43]. Reproduced with permission from [43]. Copyright 1996 John Wiley & Sons

commonly used algorithm is the filtered-x least-mean-square (FXLMS) algorithm, which was previously discussed. Figure 7.5 is an illustration of how FXLMS is applied to neutralize the noise inside a signal-channel duct-acoustic system. The noise source (reference input) is picked up by the reference microphone in the upstream, and an error microphone placed in the downstream is used to monitor the noise in the output. The reference signal $x(n)$ and error signal $e(n)$ are processed by an active noise controller implemented by the FXLMX algorithm. It generates a control signal $y(n)$ to drive a loudspeaker in order to cancel the noise.

In addition to noise control, FXLMS has been used to suppress the vibrations of composite structure [44], gear pairs [45], building [46], and machine tool [47]. FXLMS could also be combined with other controllers, such as H_∞ feedback robust controller [48] and LQR controller [49], to promote convergence speed and increase robust performance. Even though FXLMS uses a feedforward adaptive filter to change the input according to the error, it still needs off-line identification of a system as *a priori* information. However, it can be modified by adding another adaptive FIR filter to identify the system in real-time. Multiresolution analysis (MRA) can be integrated into the on-line FXLMS structure by putting analysis filter banks in front of the adaptive filters to manipulate wavelet coefficients, and then using synthesis methods to reconstruct the controlled signal. The scheme of such a construct that possesses joint time-frequency resolution and follows the on-line FXLMS algorithm is able to control nonautonomous, nonstationary systems.

7.2.1.1 Time Domain Discrete Wavelet Transform

Simultaneous time-frequency control can be realized through manipulating the discrete wavelet coefficients in the time domain [50, 51]. Implementation of the unique and novel control idea includes incorporating discrete wavelet transform (DWT) with least-mean-square (LMS) adaptive filters to perform feedforward control, on-line identification [52], and adopting a filtered-x least-mean-square (FXLMS) algorithm [43, 53] to construct parallel adaptive filter banks. In Chapter 5 it was shown that DWT in the time domain can be realized by passing the input signal through a two-channel filter bank iteratively, as the one shown in Figure 7.6.

Assume that the infinite input signal sequence $x[n]$ is of real numbers. The decomposition process in Figure 7.6 convolutes the input $x[n]$ with a high-pass filter h_0 and a low-pass filter h_1, followed by down-sampling by two. The approximation coefficient $a[n]$ and detail

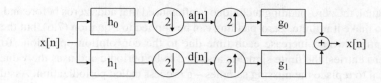

Figure 7.6   Two-channel filter bank in time domain

coefficients $d[n]$ it receives are calculated in the time domain as follows:

$$a\,[n] = \sum_k h_0\,[2n - k]\,x\,[k] = \sum_k h_0\,[k]\,x\,[2n - k] \qquad (7.4)$$

$$d\,[n] = \sum_k h_1\,[2n - k]\,x\,[k] = \sum_k h_1\,[k]\,x[2n - k] \qquad (7.5)$$

where integer $n = 0, \ldots, \infty$. Assume that the orthogonal filter sets are of equal and even length, and the lengths (or weights) of the high-pass filter h_0 and low-pass filter h_1 are both 4. Equations (7.4) and (7.5) can be carried out by multiplying the signal with a linear transformation matrix, T_a, as

$$Y = T_a X \qquad (7.6)$$

where the infinite analysis matrix, T_a, is defined below

$$T_a =$$

$$\begin{bmatrix}
\ddots & \ddots & \ddots & \ddots & \ddots & \ddots \\
\cdots & 0 & h_0[3] & h_0[2] & h_0[1] & h_0[0] & 0 & 0 & 0 & 0 & 0 & \cdots \\
\cdots & 0 & 0 & 0 & h_0[3] & h_0[2] & h_0[1] & h_0[0] & 0 & 0 & 0 & \cdots \\
\cdots & 0 & 0 & 0 & 0 & 0 & h_0[3] & h_0[2] & h_0[1] & h_0[0] & 0 & \cdots \\
& \vdots & \vdots & \vdots & \vdots & \vdots & \vdots & \vdots & \vdots & \vdots & \vdots \\
\cdots & 0 & h_1[3] & h_1[2] & h_1[1] & h_1[0] & 0 & 0 & 0 & 0 & 0 & \cdots \\
\cdots & 0 & 0 & 0 & h_1[3] & h_1[2] & h_1[1] & h_1[0] & 0 & 0 & 0 & \cdots \\
\cdots & 0 & 0 & 0 & 0 & 0 & h_1[3] & h_1[2] & h_1[1] & h_1[0] & 0 & \cdots \\
& & \ddots & & \ddots & & \ddots & & \ddots & & \ddots & \ddots
\end{bmatrix}$$

$$(7.7)$$

and X is an infinite array of the input signal. Y, which consists of wavelet approximation and detail coefficients, can be represented as

$$Y = \begin{bmatrix} \cdots & a[0] & a[1] & a[2] & \cdots & d[0] & d[1] & d[2] & \cdots \end{bmatrix}^{\mathrm{T}} \qquad (7.8)$$

For a finite signal, the zero-padding technique is often used that adds zeros before and after the finite signal to make it infinite. Zero-padding can be applied to Equation (7.6), but the number of nonzero entries would increase each time due to the convolution operation. To prevent adding nonzero entries, the finite signal is assumed to be periodic and uses the values within the finite signal to replace the missing samples – a process called periodization. Assuming the input signal X in Equation (7.6) is periodic with a period, N,

$$X = \begin{bmatrix} \cdots & x[0] & x[1] & \cdots & x[N-1] & x[0] & x[1] & \cdots & x[N-1] & \cdots \end{bmatrix}^{\mathrm{T}} \quad (7.9)$$

Equation (7.9) can be truncated as

$$X^N = \begin{bmatrix} x[0] & x[1] & \cdots & x[N-1] \end{bmatrix}^{\mathrm{T}} \quad (7.10)$$

Because the transformed signal is also periodic with period N, N consecutive entries in Y are selected to represent it. Thus a finite signal X^N is transformed into another finite signal Y^N of equal length. The analysis matrix T_a is also truncated to an $N \times N$ matrix, T_a^N, to avoid extending the signal. The deleted filter coefficient in T_a^N is put back into the proper position in the matrix to be consistent with the periodic signature of the signal. The transformation in Equation (7.6) then becomes

$$Y_{k+1}^N = T_a^N X_k^N \quad (7.11)$$

where k is the level of transformation (decomposition). For $N = 8$, the truncated analysis matrix is

$$T_a^N = \begin{bmatrix} h_0[3] & h_0[2] & h_0[1] & h_0[0] & 0 & 0 & 0 & 0 \\ 0 & 0 & h_0[3] & h_0[2] & h_0[1] & h_0[0] & 0 & 0 \\ 0 & 0 & 0 & 0 & h_0[3] & h_0[2] & h_0[1] & h_0[0] \\ h_0[1] & h_0[0] & 0 & 0 & 0 & 0 & h_0[3] & h_0[2] \\ h_1[3] & h_1[2] & h_1[1] & h_1[0] & 0 & 0 & 0 & 0 \\ 0 & 0 & h_1[3] & h_1[2] & h_1[1] & h_1[0] & 0 & 0 \\ 0 & 0 & 0 & 0 & h_1[3] & h_1[2] & h_1[1] & h_1[0] \\ h_1[1] & h_1[0] & 0 & 0 & 0 & 0 & h_1[3] & h_1[2] \end{bmatrix} \quad (7.12)$$

Substituting the above, Equation (7.11) can be rearranged to take up a concise form below

$$\begin{bmatrix} A_{k+1} \\ D_{k+1} \end{bmatrix} = T_a^N X_k^N \quad (7.13)$$

with

$$A_{k+1} = [a[0] \quad a[1] \quad \cdots \quad a[N/2-1]]^{\mathrm{T}} \quad (7.14)$$

$$D_{k+1} = [d[0] \quad d[1] \quad \cdots \quad d[N/2-1]]^{\mathrm{T}} \quad (7.15)$$

The synthesis matrix T_s^N can be constructed as

$$T_a^N T_s^N = I^N \tag{7.16}$$

meaning that

$$T_s^N = (T_a^N)^{-1} \tag{7.17}$$

and for orthogonal filters,

$$(T_a^N)^{-1} = (T_a^N)^T \tag{7.18}$$

When $k = 0$, Equation (7.13) is defined as the first level of DWT decomposition. The second level of the DWT decomposition can be expressed in the matrix form as

$$\begin{bmatrix} A_{k+2} \\ D_{k+2} \\ D_{k+1} \end{bmatrix} = \begin{bmatrix} T_a^{N/2} & 0 \\ 0 & I^{N/2} \end{bmatrix} T_a^N X_k^N \tag{7.19}$$

Similarly, the third level analysis can be implemented via

$$\begin{bmatrix} A_{k+3} \\ D_{k+3} \\ D_{k+2} \\ D_{k+1} \end{bmatrix} = \begin{bmatrix} T_a^{\frac{N}{4}} & 0 & 0 \\ 0 & I^{\frac{N}{4}} & 0 \\ 0 & 0 & I^{\frac{N}{2}} \end{bmatrix} \begin{bmatrix} T_a^{\frac{N}{2}} & 0 \\ 0 & I^{\frac{N}{2}} \end{bmatrix} T_a^N X_k^N \tag{7.20}$$

The procedure can be extended for higher levels. The corresponding synthesis can be accomplished by transposing the matrix product used in the analysis.

7.2.1.2 Integration of DWT and LMS Adaptive Filter

An alternative representation to Figure 3.1, the schematics in Figure 7.7 is one of the popular finite impulse response (FIR) filters. Given a set of N filter coefficients,

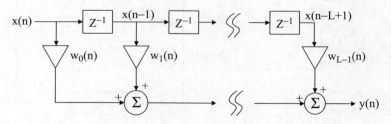

Figure 7.7 Block diagram of digital FIR filter

$w_j(n)$, $j = 0, 1, \ldots, N-1$, and a data sequence, $\{x(n), x(n-1), \ldots, x(n-N+1)\}$, the output signal is computed using

$$y(n) = \sum_{k=0}^{N-1} w_k(n) x(n-k) \tag{7.21}$$

The input vector and weight vector at time n can be defined as $X(n)$ and $W(n)$, respectively,

$$X(n) = [x(n) \quad x(n-1) \quad \ldots \quad x(n-N+1)]^{\mathrm{T}} \tag{7.22}$$

$$W(n) = [w_0(n) \quad w_1(n) \quad \ldots \quad w_{N-1}(n)]^{\mathrm{T}} \tag{7.23}$$

Consider an $N \times N$ DWT transformation matrix, T. The output signal $y(n)$ can be calculated and then compared with the desired response to determine the error signal through the following operations

$$y(n) = W^T(n) T X(n) \tag{7.24}$$

$$e(n) = d(n) - y(n) \tag{7.25}$$

The steepest-descent method is used to minimize the mean-square-error of the error signal. The least-mean-square (LMS) algorithm is used to update the weight vector as follows

$$W(n+1) = W(n) + \mu T X(n) e(n) \tag{7.26}$$

where μ is the step size for the control of stability and convergence speed. The configuration of the wavelet-based LMS adaptive filter is shown in Figure 7.8.

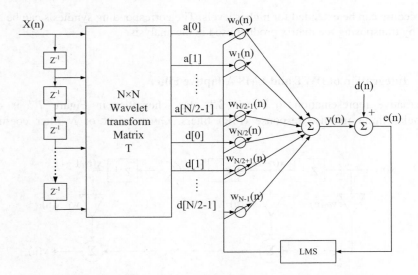

Figure 7.8   Configuration of wavelet-based LMS adaptive filter

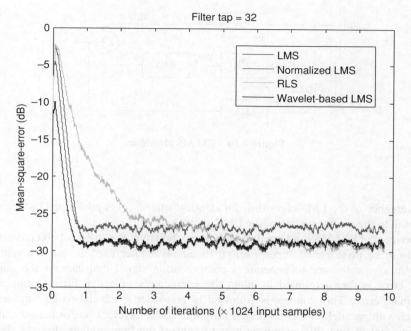

Figure 7.9   Performances of LMS, NLMS, RLS, and wavelet-based LMS

A linear dispersive channel described in Haykin's book [54] that produces unknown distortion with a random sequence input $\{x_n\}$ is used to compare the performance of the wavelet-based LMS with LMS, normalized LMS (NLMS), and recursive-least-squares (RLS). The impulse response of the channel is described by the raised cosine as follows

$$
h_n = \left\{ \begin{array}{l} \dfrac{1}{2} \left\{ 1 + \cos \left[\dfrac{2\pi}{W}(n-2) \right] \right\}, \quad n = 1,\ 2,\ 3 \\[2mm] 0, \quad otherwise \end{array} \right.
\tag{7.27}
$$

where parameter $W = 3.7$ controls the amount of amplitude distortion produced by the channel. The signal-to-noise ratio is about 30dB and the filter tap is 32. The Level-1 Daubechies-2 (db2) wavelet is employed in Figure 7.9 to show that the wavelet-based LMS is of better convergence performance.

7.2.1.3   Simultaneous Time-Frequency Control Scheme

Figure 7.10 presents an FXLMX algorithm that is a reproduction of Figure 3.3 using a slightly different set of notations. The primary path $P(z)$ defines the path from the reference source $x(n)$ to the error sensor where the noise attenuation is to be realized. The adaptive filter $W(z)$ provides an adaptive method to simultaneously model the primary plant $P(z)$ and the secondary path $S(z)$ with a given input source to minimize the residual noise $e(n)$. To ensure

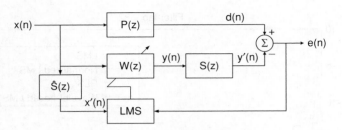

Figure 7.10   FXLMS algorithm

the convergence of the LMS algorithm, an identical filter $\hat{S}(z)$ is positioned in the reference
source path to the weight update of the LMS algorithm.

The concept of FXLMS is adopted and modified, though with a different objective. FXLMS
is based on the principle of superposition. Its adaptive filter uses the noise acquired near
the source as a reference to generate a compensating signal that cancels the noise. The
residual error is then exploited to adapt the coefficients of active filter to minimize the
mean-square-error. This concept is followed to construct a wavelet-based time-frequency
controller with parallel on-line modeling technique. The wavelet transformation matrix T is
placed in front of two FIR adaptive filters to convert the time-domain discrete signal into
a wavelet coefficient array, which is portrayed by the thick arrowed lines in the *nonlinear time-frequency control configuration* detailed in Figure 7.11. The wavelet coefficients are
multiplied by the weights of the FIR filters and then summed up to reconstruct the time-domain
signals.

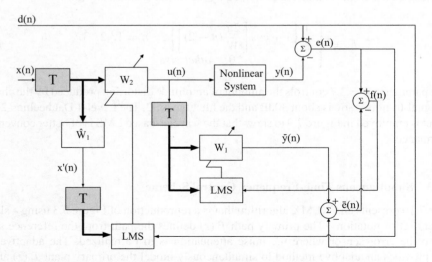

Figure 7.11   Configuration of the wavelet-based time-frequency controller

To incorporate the $N \times N$ transformation matrix T and the length-N adaptive filters, signal vectors are defined as

$$X (n) = [x(n) \quad x(n-1) \quad \cdots \quad x(n-N+1)]^T \tag{7.28}$$

$$U (n) = [u(n) \quad u(n-1) \quad \cdots \quad u(n-N+1)]^T \tag{7.29}$$

$$X' (n) = \left[x'(n) \quad x'(n-1) \quad \cdots \quad x'(n-N+1)\right]^T \tag{7.30}$$

$$E (n) = [e(n) \quad e(n-1) \quad \cdots \quad e(n-N+1)]^T \tag{7.31}$$

$$F (n) = [f(n) \quad f(n-1) \quad \cdots \quad f(n-N+1)]^T \tag{7.32}$$

These signal vectors are updated by adding the incoming data and dropping the N-th data in the array at each iteration. Thus two successive signal vectors share most of their entries. The first adaptive filter W_1 is used to model the chaotic system on-line, while the second adaptive filter W_2 serves as a feedforward controller. The weight vectors are

$$W_1 (n) = \left[w_{1,0}(n) \quad w_{1,1}(n) \quad \cdots \quad w_{1,N-1}(n)\right]^T \tag{7.33}$$

$$W_2 (n) = \left[w_{2,0}(n) \quad w_{2,1}(n) \quad \cdots \quad w_{2,N-1}(n)\right]^T \tag{7.34}$$

The identification error between the desired signal $d(n)$ and the output from W_1 can be calculated as

$$\bar{e} (n) = \bar{y} (n) - d (n) \tag{7.35}$$

where $\bar{y}(n) = W_1^T(n)TU(n)$. The error between the desired signal and the output from the chaotic system is

$$e (n) = d (n) - y (n) \tag{7.36}$$

Using Equations (7.35) and (7.36), the sequence entry can be defined

$$f (n) = e (n) - \bar{e} (n) \tag{7.37}$$

The weight of the adaptive filters is updated by the least-mean-square algorithm

$$W_1 (n+1) = W_1 (n) - \mu_1 TU (n) f (n) \tag{7.38}$$

$$W_2 (n+1) = W_2 (n) + \mu_2 TX' (n) e (n) \tag{7.39}$$

with $x' (n) = W_1^T TX(n)$ and μ_1, μ_2 being step sizes.

7.2.1.4 Optimization of Parameters

Performance of the wavelet-based time-frequency controller depends on the selection of several parameters, such as the mother wavelet and decomposition level, among others. Two

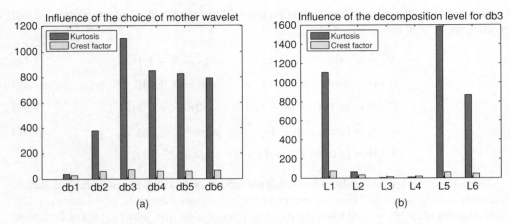

Figure 7.12   Selection of mother wavelet and decomposition level

time-domain indicators, *kurtosis* and *crest factor*, are optimized to identify the parameters that best represent the characteristics of the driving signal [55].

$$Kurtosis = \frac{\frac{1}{N}\sum_{n=1}^{N}(x(n) - \bar{x})^4}{\left[\frac{1}{N}\sum_{n=1}^{N}(x(n) - \bar{x})^2\right]^2} \tag{7.40}$$

$$Crest\ Factor = \frac{Crest\ Value}{RMS} = \frac{\sup|x(n)|}{\sqrt{\frac{1}{N}\sum_{n=1}^{N}[x(n)]^2}} \tag{7.41}$$

where $x(n)$ is the signal, N is the number of the samples, and $\bar{x}$ is the sample mean. The values of these two indicators that correspond to the Daubechies wavelet family [56] and decomposition level are shown in Figure 7.12, which indicates that the Daubechies-3 (db3) wavelet best characterizes the signal given that it has the highest kurtosis and crest factors. Even though the decomposition level 5 has the highest kurtosis value, decomposition level one is selected to ease the computational load.

When the wavelet-based time-frequency controller is applied to a system undergoing bifurcation, it is able to restrain both the time- and frequency-domain responses and keep the system in periodic motion. Hence the mitigation imposed by the controller effectively stabilizes the dynamics before it deteriorates dynamically to eventual chaos. Having the concepts of adaptive control, active noise control, and wavelet-based FIR filters all integrated, the wavelet-based time-frequency controller exerts control in the joint time-frequency domain and therefore differentiates itself from all published controllers in philosophy, architecture, and performance in mitigating nonlinear, nonstationary responses such as bifurcation and route-to-chaos.

7.3   Validation of Chaos Control

The philosophy of the proposed control scheme is that the controller must be able to inhibit the deterioration of time and frequency responses simultaneously before the frequency response is too broadband to be controlled. It employs adaptive filters for real-time system identification to cope with the nonstationary nature of the system during route-to-chaos. Additional adaptive filters are placed to adjust the input signal, compensate the emerging frequency during bifurcation, and track the reference signal. As such, the control scheme elaborated in detail in the previous section is able to regulate a chaotic system in both the time and frequency domain simultaneously. The architecture of the proposed controller adopts the active noise control algorithm (filtered-x least-mean-square) that uses one auto-adjustable finite impulse response (FIR) filter to identify the system and another auto-adjustable FIR filter to eliminate the uncontrollable input. Analysis wavelet filter banks are also incorporated. Analysis filter banks are used to decompose the input signal before entering the controller and synthesis methods to combine the control input. By projecting the input signal onto orthogonal subspaces spanned by the wavelet filter banks, the convergence performance of the least-mean-square algorithm is improved. In addition, the signal is resolved by DWT into components at various dyadic scales corresponding to successive octave frequencies, and moving wavelet filters are applied to extract temporal contents of the signal. The Daubechies orthogonal db2 wavelet is employed in the study to control the period-doubling bifurcation generated by two types of Duffing oscillators. As Daubechies wavelet functions of higher order do not provide improved time and frequency resolutions or better performance of control for the particular Duffing oscillators, the db2 wavelet is chosen for its short filter length [57]. The control law incorporating the db2 wavelet is inherently constructed in the simultaneous time-frequency space.

A double-well Duffing oscillator with nonstationary external excitation is investigated using the following system parameters: $\mu = 0.4$, $\beta = 1$, $\alpha = -0.8$, $a = 0.32$, $b = 3 \times 10^{-5}$ and $\omega = 0.78$.

$$\ddot{x} + \mu \dot{x} + \beta x + \alpha x^3 = (a + bt)\cos(\omega t) \tag{7.42}$$

The Duffing oscillator is selected for demonstrating the proposed control methodology for the reason that it exhibits period-doubling bifurcation undergoing route-to-chaos – a property shared by a broad set of nonlinear systems. An example of a periodically forced oscillator with a nonlinear elasticity, Duffing is among the most widely investigated equations. Time-frequency control of the Duffing oscillator, as established and reported in the following, provides an alternative to all the control methodologies ever documented and available in the literature.

The time response and bifurcation diagrams given in Figures 7.13(a) and (b) show that the motion is initially periodic. When the excitation amplitude is increased in time, the motion becomes a period-doubling bifurcation. When multiple periods show up, the system becomes chaotic. Bifurcation becomes prominent at this stage with an increasing but bounded frequency bandwidth. When the excitation amplitude exceeds 0.42, the response becomes unbounded. The instantaneous frequency in Figure 7.13(c) shows a dominant frequency oscillating between 0.1 and 0.15 Hz, a second frequency between 0.05 and 0.1 Hz, and a third frequency at around 0.03 Hz. They all display temporal-modal behaviors and singularities signified by spectra of infinite bandwidth. The period-doubling bifurcation initiates at $t = 2700$ sec when the dominant frequency loses its characteristics and a second component shows up with a frequency that is

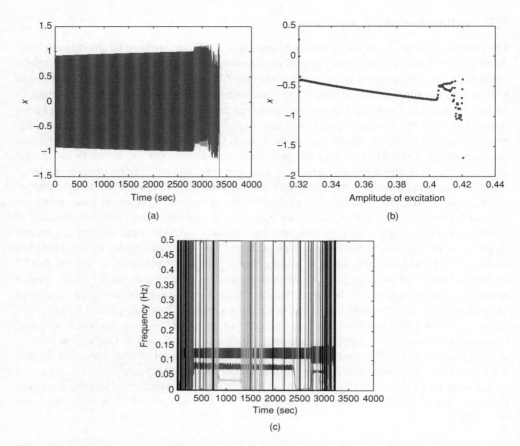

Figure 7.13 (a) Time response, (b) bifurcation diagram, and (c) instantaneous frequency of the Duffing oscillator with nonstationary external excitation. Reproduced with permission from Meng-Kun Liu, C. Steve Suh, 2012, "Simultaneous Time-Frequency Control of Bifurcation and Chaos," Communications in Nonlinear Science and Numerical and Numerical Simulation, (17) pp. 2539–2550. Copyright 2012 Elsevier

half of the dominant one. With increasing excitation amplitude, the system undergoes route-to-chaos, in which all the frequencies are seen to engage in different patterns of temporal-modal oscillations indicative of dynamic deterioration. At $t = 3300$ sec this state of instability reaches a point that renders the system no longer bounded. This phenomenon is analogous to many real-world complex nonlinear systems including the capsizing of a ship.

To demonstrate the performance of the wavelet-based time-frequency control in mitigating route-to-chaos, two scenarios are considered against the baseline case in Figure 7.13. A controller configuration is developed for this particular task. The configuration along with the implementation algorithm is schematically given in Figure 7.14. First, the controller is turned on at the beginning of the bifurcation when the corresponding spectrum is still narrowband with a finite number of temporal-modal oscillations. Second, the controller is activated in the midst of chaos when the corresponding spectrum is already broadband. The result of the first scenario

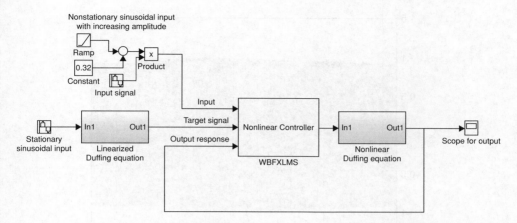

Figure 7.14   Control architecture showing implementation algorithm

is shown in Figure 7.15. Figure 7.15(a) gives the time response of the controlled signal. The controller is turned on at $t = 3000$ sec, and the amplitude of the response is quickly contained, thus a contrast to Figure 7.13(a) when no controller is applied to negate the state of route-to-chaos. The desired signal for the controller is designed by using a stationary external excitation as the input to the linearized Duffing oscillator. The error between the output and the desired signal is shown in Figure 7.15(b). The error before the onset of the controller is trivial and remains zero. The time-domain error is bounded within a small range after the controller is turned on.

Figures 7.16(a) and (b) give the bifurcation diagram and instantaneous frequency, respectively, of the controlled response. Figure 7.16(a) indicates that when the excitation amplitude is small, the response is of a single, slow-changing frequency. The controller is activated when period-doubling bifurcation just initiates. After a short transient period, the bifurcation diagram shows a rapid restoration of dynamic stability. The instantaneous frequency in Figure 7.16(b) shows that, after the transient response is stabilized, the characteristic of the dominant frequency is restored to its status before bifurcation. At the same time the singularities of infinite bandwidth are eliminated and all frequencies are now of a well-defined temporal-modal structure. Figures 7.15 and 7.16 together demonstrate that the wavelet-based time-frequency controller is able to mitigate a bifurcating system from deteriorating further in both the time and frequency domain.

The second scenario, in which the controller is turned on at the state of chaos, is presented in Figure 7.17. The scenario depicted in Figure 7.17(a) is similar to but different from Figure 7.16(a) in that the onset of the controller is postponed until $t = 3200$ sec when the system is in the state-of-chaos and on the verge of sudden divergence, as seen in Figure 7.13. The time response after the controller is engaged displays steady amplitudes. The response error in Figure 7.17(b) is bounded within a small range after a short transient period.

The bifurcation diagram and instantaneous frequency of the controlled response in Figure 7.18 corroborate the same observation that the controller is both effective and robust. The bifurcation in Figure 7.18(a) is a state of a plethora of periods and indistinguishable trajectories, and also of an increasing but finite spectrum. The wavelet-based time-frequency controller is activated at this particular moment and state of dynamic instability. After a short transient period, the trajectory in the bifurcation diagram is stabilized and becomes a well-behaving line

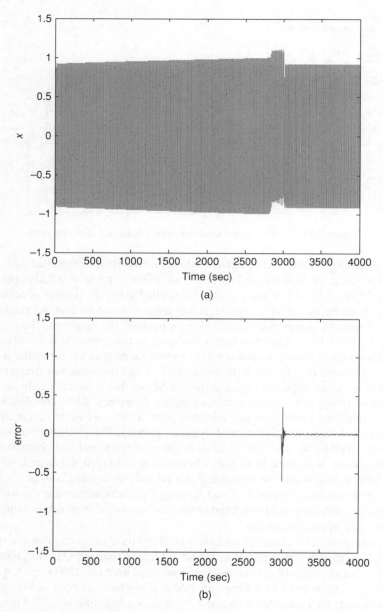

Figure 7.15 (a) Time response (b) Error response when the controller is turned on at the initial state of period-doubling bifurcation. Reproduced with permission from Meng-Kun Liu, C. Steve Suh, 2012, "Simultaneous Time-Frequency Control of Bifurcation and Chaos," Communications in Nonlinear Science and Numerical and Numerical Simulation, (17) pp. 2539–2550. Copyright 2012 Elsevier

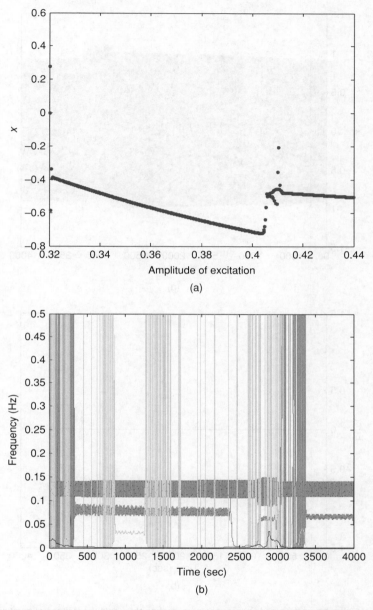

Figure 7.16 (a) Bifurcation diagram (b) Instantaneous frequency when the controller is turned on at the initial state of period-doubling bifurcation. Reproduced with permission from Meng-Kun Liu, C. Steve Suh, 2012, "Simultaneous Time-Frequency Control of Bifurcation and Chaos," Communications in Nonlinear Science and Numerical and Numerical Simulation, (17) pp. 2539–2550. Copyright 2012 Elsevier

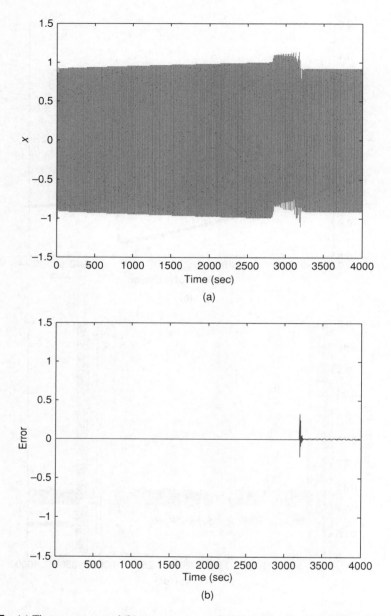

Figure 7.17   (a) Time response and (b) response error when the controller is turned on at state of chaos. Reproduced with permission from Meng-Kun Liu, C. Steve Suh, 2012, "Simultaneous Time-Frequency Control of Bifurcation and Chaos," Communications in Nonlinear Science and Numerical and Numerical Simulation, (17) pp. 2539–2550. Copyright 2012 Elsevier

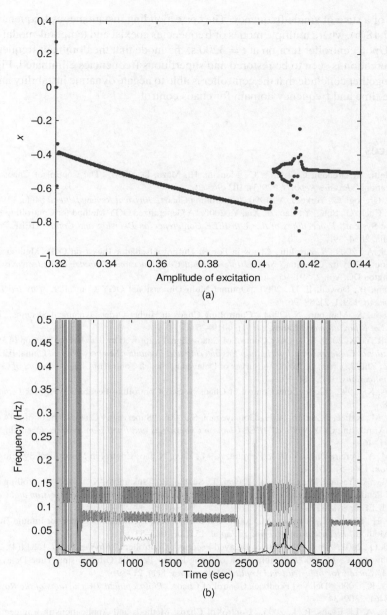

Figure 7.18 (a) Bifurcation diagram and (b) instantaneous frequency when the controller is turned on at state of chaos. Reproduced with permission from Meng-Kun Liu, C. Steve Suh, 2012, "Simultaneous Time-Frequency Control of Bifurcation and Chaos," Communications in Nonlinear Science and Numerical and Numerical Simulation, (17) pp. 2539–2550. Copyright 2012 Elsevier

indicative of a state of single frequency. The corresponding instantaneous frequency is shown in Figure 7.18(b), where multiple modes of frequencies coexist and temporal-modal aberration abounds. Upon controller turn-on at $t = 3200$ s, the mode that the dominant frequency was in before bifurcation is seen to be restored and superfluous frequencies eliminated. Figures 7.17 and 7.18 together conclude that the controller is able to negate dynamic instability and mitigate in both the time and frequency domain for chaos control.

References

[1] Boccaletti, S., Grebogi, C., Lai, Y. C., Mancini, H., Maza, D., 2000, "The Control of Chaos: Theory and Applications," *Physics Reports*, 329(3), 103–96.

[2] Ott, E., Grebogi, C., Yorke, J. A., 1990, "Controlling Chaos," *Physical Review Letters*, 64(1), 1196–99.

[3] Yu, X., Chen, G., Song, Y., Cao, Z., Xia, Y., 2000, "A Generalized OGY Method for Controlling Higher Order Chaotic System," *Proceedings of the 39th IEEE Conference on Decision and Control*, IEEE Press, Sydney, Australia, 2054–59.

[4] Sanayei, A., 2009, "Controlling Chaos in Forced Duffing Oscillator Based on OGY Method and Generalized Routh-Hurwitz Criterion," *Second International Conference on Computer and Electrical Engineering*, Washington DC, Vol. 2, 591–95.

[5] Epureanu, B., Dowell, E. H., 2000, "Optimal Multi-Dimensional OGY Controller," *Physica D: Nonlinear Phenomena*, 139(1–2), 87–96.

[6] Boukabou, A., Mansouri, N., 2004, "Controlling Chaos in Higher-Order Dynamical Systems," *International Journal of Bifurcation and Chaos*, 14(11), 4019–25.

[7] Wang, R, Yi, X, 2009, "The OGY Control of Chaos Signal Improved by The Reconstruction of Phase Space," *International Conference on Information Engineering and Computer Science*. Wuhan, China, IEEE Press

[8] Tian, Y., Zhu, J., Chen, G., 2005, "A Survey of Delayed Feedback Control of Chaos," *Journal of Control Theory and Applications*, 3(4), 311–19.

[9] Pyragas, K., 1992, "Continuous Control of Chaos by Self-Controlling Feedback," *Physics Letters A*, 170(6), 421–28.

[10] Nazari, M., Rafiee, G., Jafari, A. H., Golpayegani, S., 2008, "Supervisory Chaos Control of a Two-Link Rigid Robot Arm Using OGY Method," *IEEE Conf. on Cybernetics and Intelligent Systems*, Chengdu, China, IEEE Press 41–46.

[11] Savi, M. A., Pereira-Pinto, F. H. I., Ferreira, A. M., 2006, "Chaos Control in Mechanical Systems," *Shock and Vibration*, 13(4–5), 301–4.

[12] OGAWA, A., Yasuda, M., Ozawa, Y., Kawai, T., Suzuki, R., Tsukamoto, K., 1996, "Controlling Chaos of the Forced Pendulum with the OGY Method," *Proceedings of the 35th Conference on Decision and Control*, Kobe, Japan, IEEE Press, 2377–78.

[13] Okuno, H., Kanari, Y., Takeshita, M., 2005, "Control of Three-Synchronous-Generator Infinite-Bus System by OGY Method," *Electrical Engineering in Japan*, 151(2), 1047–53.

[14] Morgül, Ö., 2003, "On the Stability of Delayed Feedback Controllers," *Physics Letter A*, 314(4), 278–85.

[15] Postlethwaite, C. M., 2009, "Stabilization of Long-Period Periodic Orbits Using Time-Delayed Feedback Control," *Society for Industrial and Applied Mathematics*, 8(1), 21–39.

[16] Pyragas, K., 2006, "Delayed Feedback Control of Chaos," *Philosophical Transactions of the Royal Society A*, 364(1846), 2309–34.

[17] Fradkov, A. L., Evans, R. J., 2005, "Control of Chaos: Methods and Applications in Engineering," *Annual Reviews in Control*, 29(1), 33–56.

[18] Behal, A., Dixon, W., Dawson, D. M., Xian, Bin, 2010, *Lyapunov-Based Control of Robotic Systems*, Boca Raton, FL, CRC Press.

[19] Park, H. J., 2005, "Chaos Synchronization of A Chaos System via Nonlinear Control," *Chaos, Solitons and Fractals*, 25(3), 579–84.

[20] Song, L., Yang, J., 2009, "Chaos Control and Synchronization of Dynamical Model of Happiness with Fractional Order," *Industrial Electronics and Applications conference*, Xi'an, China, IEEE Press, 919–24.

[21] Daafouz, J., Millerioux, G., 2000, "Parameter Dependent Lyapunov Function for Global Chaos Synchronization of Discrete Time Hybrid Systems," *Control of Oscillations and Chaos 2nd International Conference*, St Petersburg, Russia, IEEE Press, Vol. 2, 339–42.

[22] Yang, Z., Jiang, T., Jing, Z, 2010, "Chaos Control in Duffing-Van Der Pol System," *International Workshop on Chaos-Fractal Theory and its Applications*, 106–10.

[23] Nijmeijer, H., Berghuis, H., 1995, "On Lyapunov Control of the Duffing Equation," *IEEE Transactions on Circuits and Systems I: Fundamental Theory and Applications*, 42(8), 473–77.

[24] Wang, R., Jing, Z., 2004, "Chaos Control of Chaotic Pendulum System," *Chaos, Solitons and Fractals*, 21(1), 201–7.

[25] Yazdanpanah, Am., Khaki-Sedigh, A., Yazdanpanah, Ar., 2005, "Adaptive Control of Chaos in Nonlinear Chaotic Discrete-time systems," *International Conference on Physics and Control*, 4th IEEE Conference on Industrial Electronics and Applications, Beijing, China, IEEE Press, 913–15.

[26] Bessa, W. M., Paula, A. S., Savi, M. A., 2009, "Chaos Control using an Adaptive Fuzzy Sliding Mode Controller with Application to a Nonlinear Pendulum," *Chaos, Solitons and Fractals*, 42(2), 784–91.

[27] Huang, L., Yin, Q., Sun, G., Wang, L., Fu, Y., 2008, "An Adaptive Observer-based Nonlinear Control for Chaos Synchronization," *2nd International Symposium on System and Control in Aerospace and Astronautics*, Shenzhen, China, IEEE Press, 1–4.

[28] Chen, D. Y., Ma, X. Y., 2010, "A Hyper-chaos with only One Nonlinear Term and Its Adaptive Synchronization and Control," *Control and Decision Conference*, Shenzhen, China, IEEE Press, 1689–94.

[29] Yang, B, Suh, C.S., 2004, "On The Nonlinear Features of Time-Delayed Feedback Oscillators," *Communications in Nonlinear Science and Numerical Simulations*, 9(5), 515–29.

[30] Lynch, S., 2003, *Dynamical Systems with Applications using MATLAB®*, Chapter 6, Birkhäuser, New York.

[31] Yang, B., Suh, C.S., 2003, "Interpretation of Crack Induced Nonlinear Response Using Instantaneous Frequency," *Mechanical Systems and Signal Processing*, 18(3), 491–513.

[32] Jeevan, L. G., Malik, V., 2010, "A Wavelet Based Multi-Resolution Controller," *Journal of Emerging Trends in Computing and Information Sciences*, 2(special issue), 17–21.

[33] Cloe, M. O. T., Keogh, P. S., Burrows, C. R., Sahinkaya, M. N., 2006, "Wavelet Domain Control of Rotor Vibration," *Proceedings of the Institution of Mechanical Engineers, Part:C Journal of Mechanical Engineering Science*, 220(2), 167–84.

[34] Tsotoulidis, S., Mitronikas, E. and Safacas, A., 2010, "Design of A Wavelet Multiresolution Controller for a Fuel Cell Powered Motor Drive System," *XIX International Conference on Electrical Machine*, Rome, IEEE Press.

[35] Cade, I. S., Keogh, P. S., Sahinkaya, M. N., 2007, "Rotor/active Magnetic Bearing Transient Control using Wavelet Predictive Moderation," *Journal of Sound and Vibration*, 302(1), 88–103.

[36] Yousef, H. A., Elkhatib, M. E., Sebakhy, O. A., 2010, "Wavelet Network-based Motion Control of DC Motors," *Expert System with Applications*, 37(2), 1522–27.

[37] Cruz-Tolentino, J. A., Ramos-Velasco, L. E., Espejel-Rivera, M. A., 2010, "A Self-tuning of a wavelet PID controller," *20th International Conference on Electronics, Communications and Computer*, Cholula, Mexico, IEEE Press, 73–78.

[38] Sanner, R. M., Slotine, J. E., 1995, "Structurally Dynamic Wavelet Networks for the Adaptive Control of Uncertain Robotic Systems," *Proceedings of the 34th Conference on Decision & Control*, New Orleans, LA, IEEE Press, 2460–67.

[39] Hsu, C. F., Lin, C. M., Lee, T. T., 2006, "Wavelet Adaptive Backstepping Control for a Class of Nonlinear Systems," *IEEE Transactions on Neural Networks*, 17(5), 1175–83.

[40] Lin, C. M., Hung, K. N., Hsu, C. F., 2007, "Adaptive Neuro-Wavelet Control for Switching Power Supplies," *IEEE Transactions on Power Electronics*, 22(1), 87–95.

[41] Polycarpou, M. M., Mears, M. J., Weaver, S. E., 1997, "Adaptive Wavelet Control of Nonlinear Systems," *Proceedings of the 36th Conference on Decision & Control*, San Diego, CA, IEEE Press, 3890–95.

[42] Mohammadi, S. J., Sabzeparvar, M., Karrari, M., 2010, "Aircraft Stability and Control Model using Wavelet Transforms," *Proceedings of the Institution of Mechanical Engineers, Part G: Journal of Aerospace Engineering*, 224(10), 1107–18.

[43] Kuo, S. M. and Morgan, D. R., 1996, *Active Noise Control Systems: Algorithms and DSP Implementations*, John Wiley and Sons, New York.

[44] Yang, S. M., Sheu, G. J., Liu, K. C., 2005, "Vibration Control of Composite Smart Structures by Feedforward Adaptive Filter in Digital Signal Processor," *Journal of Intelligent Material Systems and Structures*, 16(9), 773–79.

[45] Guan, Y. H., Lim, T. C., Shepard, W. S., 2005, "Experimental Study on Active Vibration Control of a Gearbox System," *Journal of Sound and Vibration*, 282(3), 713–33.

[46] Peng, F. J., Gu, M., Niemann, H. J., 2003, "Sinusoidal Reference Strategy for Adaptive Feedforward Vibration Control: Numerical Simulation and Experimental Study," *Journal of Sound and Vibration*, 256(5), 1047–61.

[47] Håkansson, L., Claesson, I., Sturesson, P. O. H., 1998, "Adaptive Feedback Control of Machine-Tool Vibration based on The Filtered-x LMS Algorithm," *International Journal of Low Frequency Noise, Vibration and Active Control*, 17(4), 199–213.

[48] Yazdanpanah, Am., Khaki-Sedigh, A., Yazdanpanah, Ar., 2005, "Adaptive control of chaos in nonlinear chaotic discrete-time systems," *International Conference on Physics and Control*, IEEE Press, 913–15.

[49] Kim, H., Adeli, H., 2004, "Hybrid Feedback-Least Mean Square Algorithm for Structural Control," *Journal of Structural Engineering*, 130(1), 120–27.

[50] Jensen, A., Cour-Harbo, A. La, 2001, *Ripples in Mathematics*, Springer-Verlag, Berlin Heidelberg.

[51] Schniter, P., 2005, "Finite-Length Sequences and the DWT Matrix," Connexions. June 9, 2005. http://cnx.org/content/m10459/2.6/.Attallah, S., 2000, "The Wavelet Transform-Domain LMS Algorithm: A More Practical Approach," *IEEE Transactions on Circuit and Systems: -II: Analog and Digital Signal Processing*, 47(3), 209–13.

[52] Kou, S. M., Wang, M., Chen, Ke, 1992, "*Active* Noise Control System with Parallel On-Lline Error Path Modeling Algorithm," *Noise Control Engineering Journal*, 39(3), 119–27.

[53] Haykin, S., 2002, *Adaptive Filter Theory*, Prentice-Hall, Upper Saddle River, NJ, Chapter 5.

[54] Chiementin, X., Kilundu, B., Dron, J. P., Dehombreux, P., Debray, K., 2010, "Effect of Cascade Methods on Vibration Defects Detection," *Journal of Vibration and Control*, 17(4), 567–77.

[55] I. Daubechies, 1998, "Orthonormal Bases of Compactly Supported Wavelets," *Communications on Pure and Applied Mathematics*, XLI (1988), 909–96.

[56] Strang, G., Nguyen, T, 1996, *Wavelets and Filter Banks*, Wellesley-Cambridge Press, Wellesley.

8

Time-Frequency Control of Milling Instability and Chatter at High Speed

Milling is a highly interrupted machining process. Milling at high speed can be dynamically unstable and chatter with aberrational tool vibrations. While its associated response is still bounded in the time domain, however, milling could become unstably broadband and chaotic in the frequency domain, inadvertently causing poor tolerance, substandard surface finish, and tool damage. In this chapter instantaneous frequency along with marginal spectra is first employed to characterize the route-to-chaos process of a nonlinear, time-delayed milling model. The novel wavelet-based time-frequency controller formulated in Chapter 7 is then explored to stabilize the nonlinear response of the milling tool in the time and frequency domains concurrently. By exerting proper mitigation schemes to both the time and frequency responses, the controller effectively denies milling chatter and restores milling stability as a limit cycle of extremely low tool vibrations. The study also serves to advocate for marginal spectra as the tool of choice over Fourier spectra in identifying milling stability boundary.

8.1 Milling Control Issues

Milling is a machining operation whose high cutting efficiency is facilitated through the simple deployment of small tools of a finite of cutting edges at high spindle speed. When the immersion rate is low and the time spent cutting is only a small fraction of the spindle period, interrupted cutting would ensue as a result. The regenerative effect could also be prominent, where the cutting force depends on the current as well as the delayed tool positions. In the stability analysis performed using a linear high speed milling model, Davies *et al.* [1] showed that the fixed point of the model can lose its stability through either Neimar–Sacker bifurcation or period-doubling bifurcation. Szalai *et al.* [2] further established that both bifurcations were subcritical using a nonlinear discrete model. They also demonstrated that a stable cutting can

Control of Cutting Vibration and Machining Instability: A Time-Frequency Approach for Precision, Micro and Nano Machining, First Edition. C. Steve Suh and Meng-Kun Liu.
© 2013 John Wiley & Sons, Ltd. Published 2013 by John Wiley & Sons, Ltd.

suddenly turn into chatter – a pronounced dynamic effect characterized by large tool vibration amplitude or frequency oscillation different from the spindle speed. Such a negative effect induces detrimental aperiodic errors, such as waviness on the workpiece surface, inaccurate dimensions, and excessive tool wear, among others [3].

The onset of chatter has been investigated both analytically and numerically. Dynamic milling equations were transformed into linear maps and the eigenvalues of the transition matrix in the complex plane were used to predict stability [1, 2]. Using numerical integration, stability was predicted by gradually increasing the axial depth-of-cut until instability occurred [3]. However, each method has its own shortcomings. Established methodologies use eigenvalues of the approximated transition matrix to determine the stability bound of the system. In the route-to-chaos process, the way these eigenvalues leave the unit circle in the complex plane is used to identify the type of bifurcation. But as long as the high order nonlinear terms are omitted and the solution is projected into orthogonal eigenvectors, the response is obscured and cannot be considered a genuine representation of the nonlinear system. In numerical study, the stability of the system is decided by the emergence of additional frequencies in the corresponding Fourier spectrum. As a mathematical averaging scheme in the infinite integral sense, Fourier transform generates spectra that are misinterpreted and fictitious frequency components that are nonphysical [4]. Thus, stability determined by Fourier spectra would necessarily be erroneous. It has been demonstrated that to characterize a route-to-chaos process, both time and frequency responses need to be considered [5]. The concept of instantaneous frequency (IF) [6] introduced previously is adopted in this chapter to help manifest the dependency of frequency on time – an attribute common to all nonlinear responses, including milling chatter.

In general, contemporary control theories are developed either in the frequency domain or time domain alone. When a controller is designed in the frequency domain, the equation of motion is converted into a transfer function. Frequency response design methods, such as Bode plot and root locus, can be used to help develop frequency-domain-based controllers [7]. When a controller is designed in the time domain, the differential equations of the system are described as a state space model by state variables. Once controllability and observability are established, time domain control laws can then be applied. Controllers of either construct can only be applied exclusively either in the frequency or time domain, and they have been shown to be suitable for linear, stationary systems. However, for a nonlinear, nonstationary system, when undergoing bifurcation to eventual chaos, its time response is no longer periodic and a broadband frequency spectrum emerges. Controllers designed in the time domain confine the time error while being unable to suppress the expanding spectrum. On the other hand, controllers designed in the frequency domain constrain the frequency bandwidth while losing control over time-domain error. Neither frequency-domain nor time-domain-based controllers are sufficient to deal with bifurcation and chaotic response. This is also ascertained by the Uncertainty Principle, which states that time and frequency resolutions cannot be simultaneously achieved.

In the sections that follow, a high-speed, low immersion milling model is explored without linearization so as to retain the inherent physical attributes of the nonlinear system. Because neither linearization nor eigenvectors are attempted, tools commonly adopted for identifying various types of bifurcations are no longer applicable. As an alternative, instantaneous frequency is deployed to characterize the route-to-chaos process in the simultaneous time-frequency domain. The novel wavelet-based time-frequency controller presented in Chapter 7, along with its fundamental features that enable simultaneous time-frequency control, is also utilized. The wavelet-based controller owes its inspiration to active noise control [8], though of a different objective. While active noise controls serve to minimize acoustic noise, the wavelet-based controller is configured to mitigate the deterioration of the aperiodic response in both time and frequency domains when the system undergoes dynamic instability including bifurcation to chaos. The most prominent property of the controller is its applicability to nonlinear systems whose responses are nonautonomous and nonstationary. Such a powerful attribute is made possible by incorporating adaptive filters, so that system identification can be executed in real-time and control law can be timely modified according to the changing circumstances. Chapter 7 is referred to for a detailed account of all the components of the wavelet-based time-frequency controller, including discrete wavelet transform (DWT) in the time domain, the wavelet-based finite impulse response (FIR) filter, and the filtered-x least-mean-square (FXLMS) algorithm.

8.2 High-Speed Low Immersion Milling Model

The one degree of freedom milling model found in Ref. [2], that governs the tool motion of the cutting operation at high speed is adopted, as shown in Figure 8.1. The tool has an even number of edges and operates at a constant angular velocity, Ω. Its mass, damping coefficient, and spring coefficient are denoted as m, c, and k, respectively. The feed rate is provided by the workpiece velocity V_0. The dynamic equation of milling motion that corresponds to Figure 8.1(b) is

$$\ddot{x}(t) + 2\xi\omega_n\dot{x}(t) + \omega_n^2 x(t) = \frac{\delta(t)}{m} F_c(h(t)) \tag{8.1}$$

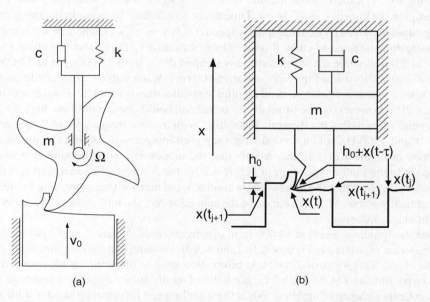

Figure 8.1 (a) Configuration and (b) mechanical model of high-speed milling

where $x(t)$ is the tool tip vertical position, $\omega_n = \sqrt{c/m}$ is the undamped natural frequency, and $\xi = c/(2m\omega_n)$ is the relative damping factor. F_c, the nonlinear cutting force, is derived from the empirical three-quarter rule as a function of the workpiece thickness, $h(t)$ [2],

$$F_c(h(t)) = Kw\,[h(t)]^{\frac{3}{4}} \tag{8.2}$$

where K is an empirical parameter and w is chip width. $h(t)$ equals the feed per cutting period h_0 plus the previous tool tip position, $x(t - \tau)$, and minus the current tool tip position, $x(t)$,

$$h(t) = h_0 + x(t - \tau) - x(t) \tag{8.3}$$

with $\delta(t)$ being a delta function defined as

$$\delta(t) = \begin{cases} 0 & if \quad \exists\, j \in \mathbb{Z}: \quad t_j \le t < t_{j+1}^- \\ 1 & if \quad \exists\, j \in \mathbb{Z}: \quad t_{j+1}^- \le t < t_{j+1} \end{cases} \tag{8.4}$$

The cutting force is applied to the system only when the tool edge physically engages the workpiece ($t_{j+1}^- \le t < t_{j+1}$). After the tool edge disengages the workpiece, the tool starts free vibration until the next edge arrives ($t_j \le t < t_{j+1}^-$). As is noted in [2], the time spent on cutting is relatively small compared to the time spent on free vibration.

8.3   Route-to-Chaos and Milling Instability

Following Ref. [1], the mass of the tool m is 0.0431 kg, stiffness coefficient k is 1.4 MN/m and damping coefficient c is 8.2 Ns/m. Time delay τ is defined by considering the number of cutting edges deployed (N) and the spindle speed (Ω) as $\tau = 2\pi/N\Omega$, where $N = 2$ in the study. In investigating the route-to-chaos displayed by Equation (8.1), the axial depth-of-cut (ADOC) is kept at 1.0 mm while the spindle speed is stepped down from 15 000 rpm to 12 000 rpm. Figure 8.2(a) shows the milling response at 15 000 rpm. It is a stable cutting condition having a time response that oscillates with an amplitude smaller than 0.1 mm. Its Fourier spectrum in Figure 8.2(b), however, is one of relatively broad bandwidth having multiple high frequency components, thus indicating dynamic instability. With a major frequency oscillating between 1000 Hz and 1300 Hz, and a second frequency oscillating about 500 Hz, the instantaneous frequency in Figure 8.2(c) further asserts that the motion is a period-doubling bifurcation. As oppose to the Fourier spectrum in Figure 8.2(b), the corresponding marginal spectrum in Figure 8.2(d) shows that the bandwidth is confined and narrow, thus signifying the response as dynamically stable. With a decrease of the spindle speed, the milling response undergoes a route-to-chaos process.

When the spindle speed is at 14 000 rpm, new frequencies that are $1/2$ and $1/4$ of the 500 Hz component are registered in Figures 8.3(c) and 8.3(d), meaning that the response is bifurcating towards chaos. This 4T period-doubling bifurcation state of instability is not resolved in the Fourier spectrum in Figure 8.3(b). Figure 8.3(b) literally misinterprets the response as one that is not experiencing bifurcation. When the spindle speed is further reduced to 13 000 rpm, the Fourier spectrum in Figure 8.4(b) remains almost unchanged from the previous one with

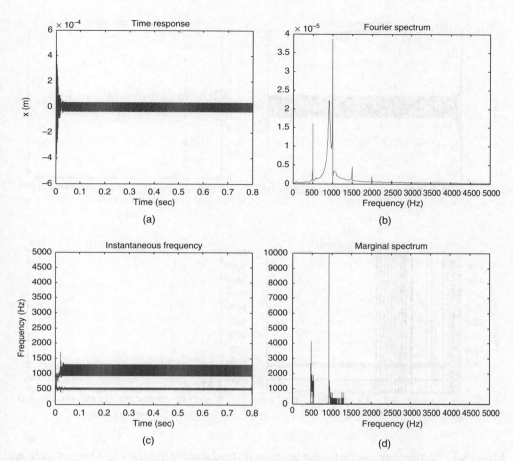

Figure 8.2 (a) Time response (b) Fourier Spectrum (c) Instantaneous frequency (d) Marginal spectrum when $\Omega = 15\,000$ rpm and ADOC $= 1.0$ mm (stable cutting condition)

an unmistakably lower time response amplitude. The corresponding instantaneous frequency and marginal spectrum in Figures 8.4(c) and 8.4(d), however, depict a scenario in which the system response is engaging in a state of very different temporal–spectral structure. With the expanding frequency bandwidth, the system is on the verge to becoming chaotic.

The Fourier spectrum in Figure 8.5(b) that corresponds to a lower spindle speed at 12 000 rpm is neither intuitively nor physically correct. The instantaneous frequency in Figure 8.5(c) shows that the major frequency oscillates between 1000 Hz and 3000 Hz, a character of a chaotic motion. This is further confirmed by the marginal spectrum in Figure 8.5(d) where a plethora of frequency components constitute a broadband spectrum. The lesson learned from Figures 8.2–8.5 is that, while the Fourier spectrum misinterprets the state of the response, instantaneous frequency along with the marginal spectrum is preferred for resolving the route-to-chaos process and deterioration of dynamic stability.

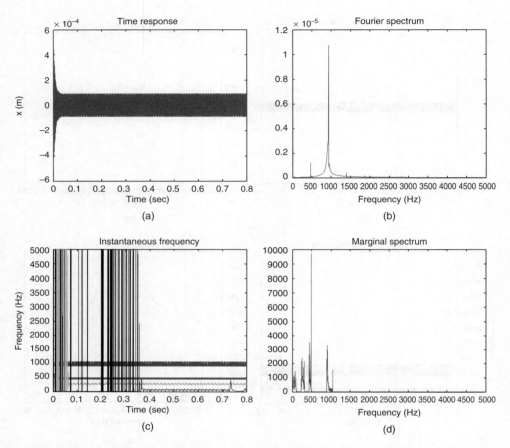

Figure 8.3   (a) Time response (b) Fourier Spectrum (c) Instantaneous frequency (d) Marginal spectrum when $\Omega = 14\,000$ rpm and ADOC $= 1.0$ mm (4T period-doubling bifurcation)

8.4   Milling Instability Control

When milling at high speed, the corresponding time response becomes aperiodic and the frequency response deviates away from well-defined harmonics and becomes unstably broadband. Such responses are highly nonlinear and could lead to tool chatter if not quickly and properly attended to. The wavelet-based time-frequency control is applied to control the time-delayed milling model in Equation (8.1) in response to a 50 000 rpm spindle speed and two different ADOCs at 3 mm and 1 mm. The schematic in Figure 8.6 gives the corresponding controller architecture that is realized and implemented as a MATLAB® Simulink® algorithm. Given that Fourier spectra have been shown to obscure the genuine characteristics of all the responses considered in the previous sections, instantaneous frequency and marginal spectrum are adopted instead. The orthogonal Daubechies-2 (db2) wavelet is once again selected as the mother wavelet in the wavelet-based controller. The filter length of the identifying filter and the controlling filter are both 256.

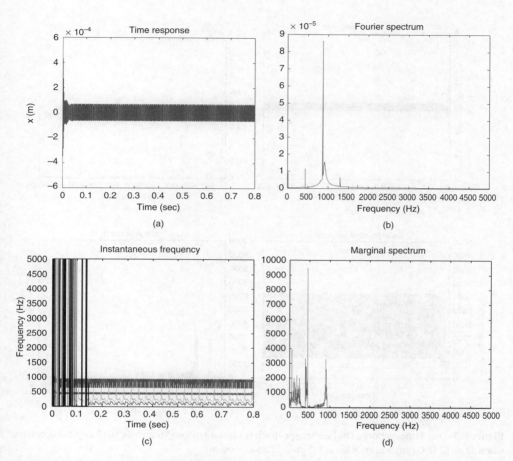

Figure 8.4 (a) Time response (b) Fourier Spectrum (c) Instantaneous frequency (d) Marginal spectrum when $\Omega = 13\ 000$ rpm and ADOC $= 1.0$ mm (unstable cutting condition)

At ADOC $= 3$ mm the wavelet-based controller is turned on at $t = 0.2$ sec to align the response with zero, the target trajectory designated for the tool. When $t \leq 0.2$ s, the vibration amplitude in Figure 8.7(a) is aberrational. There are four distinct instantaneous frequencies in Figure 8.7(c), each oscillating with the temporal-modal structure typical of a highly bifurcated response. Before the controller is applied, the marginal spectrum in Figure 8.7(b) sees a spectrum with frequencies ranging from 0 to 1500 Hz. When the controller is on-line at $t = 0.2$ s, the time response is greatly reduced and the frequency response is a well-behaved temporal-modal narrowband structure confined between 1400 Hz and 2200 Hz. As a further verification, the phase plots for the before and after scenarios are compared in Figure 8.8. The response in the state space is seen to reduce to a manifold after the controller is applied, thus explaining the restraining of bandwidth and frequencies. It can be concluded from the fact that these two phase plots belong to different basins and have fundamentally different geometric structures

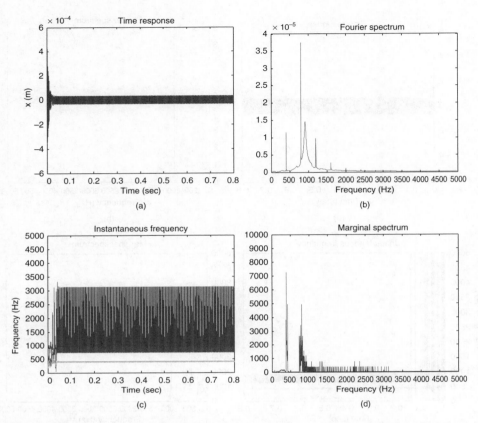

Figure 8.5 (a) Time response (b) Fourier spectrum (c) Instantaneous frequency (d) Marginal spectrum when $\Omega = 12\,000$ rpm and ADOC $= 1.0$ mm (chaotic motion)

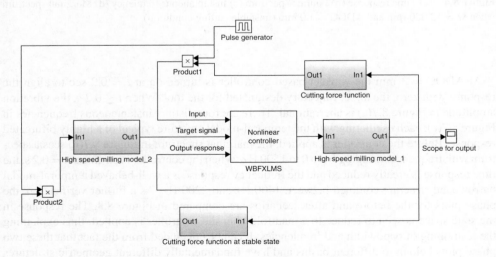

Figure 8.6 High speed milling control algorithm

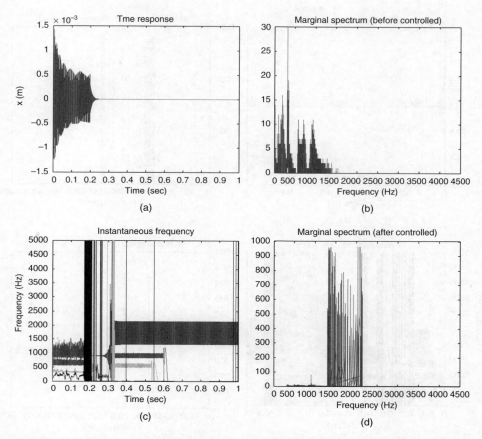

Figure 8.7 (a) Time response (b) Marginal spectrum (before controlled) (c) IF (d) Marginal spectrum (after controlled) when controller applied at $t = 0.2$ s with $\Omega = 50\,000$ rpm and ADOC = 3 mm

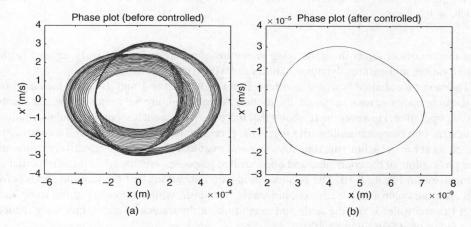

Figure 8.8 Phase plots of (a) uncontrolled and (b) controlled responses

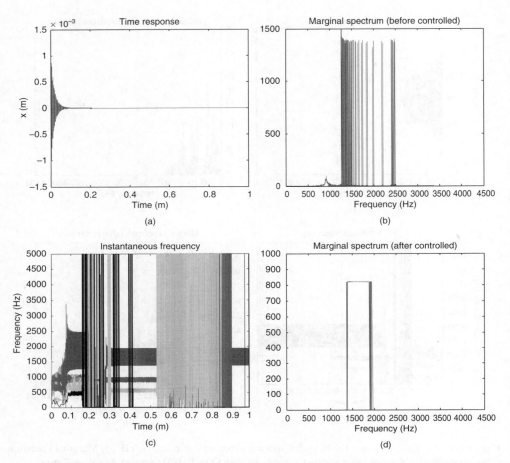

Figure 8.9 (a) Time response (b) Marginal spectrum (before controlled) (c) Instantaneous frequency (d) marginal spectrum (after controlled) when controller applied at $t = 0.2$ s with $\Omega = 50\,000$ rpm and ADOC $= 1$ mm

that the controller alters the underlying signature of the system, effectively negates further deterioration, and ensures dynamic stability at 50 000 rpm.

The next case studied is with a smaller milling ADOC set at 1 mm. Because the amplitude of the time response remains small, the motion as seen in Figure 8.9 seems to suggest a stable cutting operation. However the IF shows that there are prominent temporal-modal oscillations, indicative of a complex nonlinearity that is the precursor of tool chatter. After the controller is turned on at $t = 0.2$ s, both the time response and marginal spectrum are effectively restrained. The phase plots of the controlled and uncontrolled responses are placed next to each other for comparison in Figure 8.10. The two manifolds are indications that the motions, both before and after the controller is activated, are stable. However, while it remains in the same basin after the controller is on, the scale and magnitude of the trajectory are significantly reduced by a factor of 100 in Figure 8.10(b).

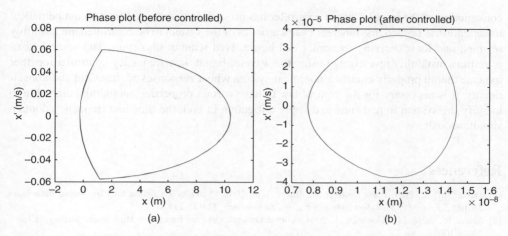

Figure 8.10   Phase plots of (a) uncontrolled and (b) controlled responses

A few observations can be made with regard to the figures above. First the wavelet-based time-frequency controller not only reduces the aberrational vibrations in the time domain. It also simultaneously regulates the spectral response in the frequency domain. Thus the nonlinearity of the response is mitigated from further deterioration. Second, the motion of the milling tool transforms dramatically after the controller is applied, as a low-amplitude oscillation in the time domain that manifests a reduction manifold in the state space. Even though the control target is set to zero and the time domain error achieved is satisfactorily small, however, the spectral response does not contain just one single frequency. The frequency spectrum still varies, but is now strictly confined within a limited bandwidth.

8.5   Summary

Milling tool dynamics was shown using instantaneous frequency, in lieu of Fourier spectra, to be transient and nonlinear due to the regenerative effect. Milling response was seen to be highly sensitive to machining condition and external perturbation, easily deteriorating from bifurcation to chaos. When losing stability, milling time response was no longer periodic and the frequency response became broadband, rendering tool chatter and tool damage probable. The marginal spectrum derived from instantaneous frequency was considered to be more suitable than Fourier spectra to define the stability boundary for high-speed milling operations. For the route-to-chaos process in which both time and frequency responses deteriorate at the same time, it is necessary to control them simultaneously. The wavelet-based time-frequency controller having DWT, the wavelet-based adaptive FIR filter, and the FXLMS algorithm as its physical features was demonstrated to successfully negate bifurcations and chaotic responses by adjusting the input. Integration of DWT in the controller effectively manipulated the wavelet coefficients, hence facilitating the control of milling tool response in both the time and frequency domains concurrently. The concept adopted from the FXLMS algorithm enabled the identification and control of the system in *real-time*. Unlike the conventional control law design approaches that always require the system to be controlled to be mathematically explicit, the construction of the wavelet-based controller has no such requirement. Since no mathematical linearization is needed, the inherent characteristics of the system to be controlled are faithfully retained and its underlying dynamics can be resolved without distortion. The several cases of milling instability investigated using the wavelet-based time-frequency controller together indicate that to properly control a nonlinear system whose responses are transient and nonstationary, it is necessary for the control law to have certain properties, including being able to identify the system in real-time and apply mitigation in both the time and frequency domain simultaneously.

References

[1] Davies, M. A., Pratt, J. R., Dutterer, B., Burns, T. J., 2002, "Stability Prediction for Low Radial Immersion Milling," *Journal of Manufacturing Science and Engineering*, 124(2), 217.

[2] Szalai, R., Stépán, G., Hogan, S. J., 2004, "Global Dynamics of Low Immersion High-Speed Milling," *Chaos*, 14(4), 1069.

[3] Balachandran, B., 2001, "Nonlinear Dynamics of Milling Process," *Philosophical Transactions of the Royal Society A*, 359(1781), 793–819.

[4] Yang, B., Suh, C.S., 2003, "Interpretation of Crack Induced Nonlinear Response Using Instantaneous Frequency," *Mechanical Systems and Signal Processing*, 18(3), 491–513.

[5] Liu, M.-K., Suh, C.S., 2012, "Temporal and Spectral Responses of A Softening Duffing Oscillator Undergoing Route-To-Chaos, *Communications in Nonlinear Science and Numerical Simulations*, 17(6), 2539–50.

[6] Huang, N.E., Shen, Z., Long, S. R., *et al.*, 1998, "The Empirical Mode Decomposition And The Hilbert Spectrum for Nonlinear And Non-Stationary Time Series Analysis," *Proceedings of the Royal Society A: Mathematical, Physical and Engineering Sciences*, 454(1971), 903–95.

[7] Franklin, G. F., Powel, J. D., Emami-Naeini, A., 1994, *Feedback Control of Dynamic Systems*, 3rd Edition, Addison-Wesley, Reading, MA.

[8] Kuo, S. M., Morgan, D.R., 1996, *Active Noise Control Systems: Algorithms and DSP Implementations*, John Wiley and Sons, New York.

9

Multidimensional Time-Frequency Control of Micro-Milling Instability

Micro-milling is inherently unstable and chattering with aberrational tool vibrations. While time response is bounded, however, micro-milling can become unstably broadband and chaotic in the frequency domain, inadvertently resulting in poor tolerance and frequent tool damage. In the chapter we apply the novel wavelet-based time-frequency control to negate the various nonlinear dynamic instabilities, including tool chatter and tool resonance, displayed by a multidimensional, time-delayed micro-milling model. We note that the time and frequency responses of the force and vibration of the model agree well with published experimental results. The multivariable control scheme formulated for study is realized by implementing two independent wavelet-based controllers in parallel to follow target signals representing the desired micro-milling state of stability. The control of unstable cutting at high spindle speeds ranging from 63 000 to 180 000 rpm and different axial depth-of-cuts is investigated using phase portrait, Poincaré map, and instantaneous frequency. The wavelet-based time-frequency control scheme effectively restores dynamic instabilities, including repelling manifold and chaotic response back to an attracting limit cycle or periodic motion of reduced vibration amplitude and frequency response. The force magnitude of the dynamically unstable cutting process is also reduced to the range of stable cutting.

9.1 Micro-Milling Control Issues

Essential for production of complex three-dimensional products from a wide range of materials, micro-milling is critical to the advancement of technology for many industries as components are continuously being reduced in size and require increased functionality. However, micro-milling is subject to unpredictable tool life and premature tool failure, which can ruin a

Control of Cutting Vibration and Machining Instability: A Time-Frequency Approach for Precision, Micro and Nano Machining, First Edition. C. Steve Suh and Meng-Kun Liu.
© 2013 John Wiley & Sons, Ltd. Published 2013 by John Wiley & Sons, Ltd.

workpiece and instigate costly and inefficient product inspection and resetting [1, 2]. Thus it would be of direct impact to improve the efficiency of the process. Chip clogging, fatigue, and excessive stress-related failure are identified as the three common micro-mill breakage mechanisms [1]. When the stress is below the endurance limit but above the normal operation level the tool will not fail immediately [1]. However, the stress on the shaft will change repeatedly while the tool is rotating causing the strain distribution to change repeatedly at the tool shaft, thus inducing fatigue. Vibrations with high or multiple frequency components increase the speed at which the strain distribution changes, inevitably resulting in fatigue failure occurring at an accelerated rate. The excessive stress-related breakage occurs when there is a sudden increase in the cutting forces indicative of dynamically unstable cutting due to excessive vibration magnitudes. Also, excessive machining vibrations (chatter) affect the workpiece surface finish and tolerances, and result in larger cutting forces, which are key indicators of tool performance [3]. Thus, micro-milling performance and failure are directly affected by the dynamic response of the tool, rendering controlling dynamic instability fundamental to improving micro-milling efficiency.

Physical models are important for the characterization of dynamic instability, development and testing of control algorithms, and providing insight necessary for designing empirical research. Micro-milling cannot directly adopt the methods used for modeling macro-milling due to different cutting force mechanisms at work, such as the increased impact of material plowing. When the chip thickness is too small a chip will not form and the material will be plowed under the tool [4]. This phenomenon is more prominent in micro-milling due to the increased feed-rate to tool nose radius ratio. Micro-milling is a highly nonlinear process due to these additional nonlinear characteristics and the high spindle speeds which are commonly employed. To address the issue of micro-milling chatter an uncut chip thickness model is coupled with a finite element orthogonal cutting model in [5]. Stability lobes are generated using statistical variances and chatter is defined as a statistical variance larger than 1 μm. However, the uncut chip thickness model reported in [5] fails to consider the elastic recovery of the material due to the plowing mechanism, thus hindering its veracity. Micro-milling stability lobes are produced in [6, 7] but the stability lobes have limited accuracy when compared with the experimental data, and it is shown that the dynamic properties of the system have a substantial impact on the resulting stability. These stability lobes are generated through linearization which obscures the nonlinear characteristics of the process that are prominent in micro-milling. The experimental data in [6, 7] show high frequency components and multiple chatter frequencies that are characteristics of a nonlinear process. The modeling and control analysis of the process should retain the inherent nonlinearities to effectively address dynamic instability. An effective method for modeling the nonlinear forcing mechanism of the micro-milling process is through slip-line field models. It is shown in [8] that the comprehensive slip-line field model developed in [9] outperforms the finite element model when predicting the magnitude of the cutting forces. A comprehensive slip-line model is developed in [10] for modeling the cutting process near the tool edge. Earlier slip-line models in [11, 12] predicted the shearing and plowing forces, and the force mechanism equations were improved upon in [13]. The research reviewed above focuses on the development of force mechanisms for predicting cutting forces and does not investigate dynamic instability. Reference [14] adopts the slip-line field force mechanism presented in [13] and accounts for material elastic recovery in the chip thickness calculation, the effective rake angle, and the helical angle of the tool for numerically studying the dynamic response. This model captures all the prominent nonlinear characteristics of micro-milling and will be adopted in this investigation to explore nonlinear micro-milling control.

There are several challenges in controlling micro-machining. In addition to the distinct cutting dynamics which differentiate micro-machining from macro-machining, the performance of the miniaturized end-mill is greatly affected by the vibrations and excessive force. The influence of noise and the inadequate bandwidth of force sensors due to high rotational speeds make it difficult to measure cutting forces [8]. Unlike macro-machining, impulse hammer tests for investigating tool tip dynamics are not practical in microstructure due to the fatigue nature of miniature tools, and the accelerometers cannot be effectively attached due to their size and weight, which influence the overall dynamics [15]. To estimate the microstructure tool dynamics, receptance coupling (RC) was implemented to mathematically couple the tool tip and the remainder of the tool using non-contact sensors [16]. However, the development of micro-machining controllers still suffers from the challenges of miniature micro-structure. There are only a few research papers related to the control of micro-machining. Command shaping method was followed to reduce the tracking error of the micro-mills in [17]. It properly chose the acceleration profile of the DC motors on the precision linear stage to counteract the vibration caused by the internal force from high speed motion of the tool. The contour error was reduced by cross-coupling, which established the real-time contour error model and returned an error correction signal to the motor of each axis [18]. Piezoelectric stack actuators were mounted to directly control the relative motion between the tool-spindle and the workpiece of a micro-milling machine, and active vibration control (AVC) was used to suppress the vibration of the tool tip point [19]. As the response time of piezoelectric actuators cannot catch up with the high rotating speed of the spindle, the method can only apply to low-speed micro-machining.

From the above, it is concluded that there are no effective solutions for controlling chatter in micro-machining. One crucial factor is to regulate the highly nonlinear cutting forces. The cutting forces cannot exceed the critical limit of the tool at which sudden tool failure would occur. This limits the chip load and thus the material removal rate that can be achieved. Higher spindle speeds are then desired to increase the material removal rates of the process without increasing the chip load. However, the nonlinearities of the force mechanism become more prominent at higher spindle speeds, causing the increased excitation frequencies to result in dynamic instability. Another factor is that the control method has to adapt to the uncertainties of the cutting process as well as the changing dynamics. However, cutting instability in fact consists of the deterioration in both time and frequency domains due to the highly nonlinear nature of micro-milling process.

This chapter is organized to address the issues identified above by first giving a detailed account of a nonlinear micro-milling model that captures the intrinsic characteristics of the cutting process. Deriving from the wavelet-based time-frequency controller design discussed in the previous chapter, a multivariable nonlinear control scheme is developed to facilitate the proper mitigation of micro-milling instability.

9.2 Nonlinear Micro-Milling Model

The micro-milling model to be controlled is one that accounts for the prominent nonlinear characteristics of the process [14]. The forcing mechanism for the model adopts the slip-line force model developed in [13], which expanded upon the model in [12] by accounting for the dead metal cap and adding an additional slip-line on the clearance face of the tool. The model

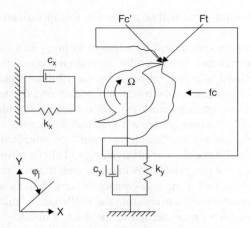

Figure 9.1 The 2D lumped mass, spring, damper model of the micro-tool. Reproduced with permission from Steve C. Suh, Meng-Kun Liu, 2013, "Multi-Dimensional Time-Frequency Control of Micro-Milling Instability," Journal of Vibration and Control. Copyright 2013 Sage

in [14] neglects this additional slip-line assuming that, since the material takes time to recover and the feed rates used are larger than the tool nose radius, this additional slip-line will have a negligible effect on the cutting forces. When the chip thickness is greater than the critical chip thickness, it is assumed that both shearing and plowing forces are present. The shearing and plowing forces in the cutting and thrust directions as given in [12, 13, 14] are provided in Equations (9.1)–(9.4)

$$dF_{sc} = \tau \, da \, [(\cos \varphi_S + a_\theta \, \sin \varphi_S) \, l_S + (\cos(2\eta_2) \sin \alpha_e + a_2 \cos \alpha_e) \, l_b] \qquad (9.1)$$

$$dF_{st} = \tau \, da \, [(a_\theta \cos \varphi_S - \sin \varphi_S) \, l_S + (\cos(2\eta_2) \cos \alpha_e - a_2 \sin \alpha_e) \, l_b] \qquad (9.2)$$

$$dF_{pc} = \tau \, da[(\cos(2\eta_1) \cos \psi + a_1 \sin \psi) \, l_b] \qquad (9.3)$$

$$dF_{pt} = \tau \, da \, [(a_1 \cos \psi - \cos(2\eta_1) \sin \psi) \, l_b] \qquad (9.4)$$

where, referring also to Figure 9.1, τ is the material shear flow stress, da is the axially depth of cut, φ_s is the chip flow angle, α_e is the effective rake angle, and l_s, l_b, a_θ, a_1, a_2, η_1, η_2, and ψ are the slip-line field variables as defined in [13, 20]. The force equations and associated variables in Equations (9.1)–(9.4) are all functions of the instantaneous chip thickness, $tc(t)$. In micro-milling, when $tc(t)$ is less than the minimum chip thickness, tc_{min}, then only plowing forces are present, and when the tool jumps out of the cut, there are no forces acting on the system. Thus, the three force cases considered are:

Case I: $tc(t) > tc_{min}$ $\quad \begin{cases} dF_t = dF_{st} + dF_{pt} \\ dF_c = dF_{sc} + dF_{pc} \end{cases}$

Case II: $0 < tc(t) < tc_{min}$ $\quad \begin{cases} dF_t = dF_{pt} \\ dF_c = dF_{pc} \end{cases}$

Case III: $tc(t) < 0$ $\quad \begin{cases} dF_t = 0 \\ dF_c = 0 \end{cases}$

These three cases indicate that accurately determining $tc(t)$ is important in order to faithfully realize the forces acting on the system and thus the resulting dynamics. The model utilizes a method which accounts for the elastic recovery of the plowing phenomenon and the tool jumping out of the cut. Equations (9.5)–(9.7) are utilized for determining $tc(t)$ where subscript, j, refers to the tooth 1 and 2 of the micro-mill, fc is the feed rate, N is the number of teeth, Ω is the spindle speed, $\Delta x = x(t) - x(t - \delta)$, $\Delta y = y(t) - y(t - \delta)$, and time delay $\delta = {}^{60}\!/_{(N\Omega)}$.

$$tc_j(t - \delta) > tc_{min}: \quad tc_j(t) = fc \cdot \sin \varphi_j(t) + \Delta x \sin \varphi_j(t) + \Delta y \cos \varphi_j(t) \quad (9.5)$$

$$0 < tc_j(t - \delta) < tc_{min}: \quad tc_j(t) = fc \cdot \sin \varphi_j(t) + \Delta x \sin \varphi_j(t) + \Delta y \cos \varphi_j(t)$$
$$+ p_e \cdot tc_j(t - \delta) \quad (9.6)$$

$$tc_j(t - \delta) < 0: \quad tc_j(t) = fc \cdot \sin \varphi_j(t) + \Delta x \sin \varphi_j(t) + \Delta y \cos \varphi_j(t)$$
$$+ tc_j(t - \delta) \quad (9.7)$$

The model also accounts for the effective rake angle and the helical angle. The derivation for the helical angle results in the following equations of force components

$$F_x = -F_t \sin \varphi_j(t) + F'_c \cos \varphi_j(t) \quad (9.8)$$

$$F_y = -F_t \cos \varphi_j(t) - F'_c \sin \varphi_j(t) \quad (9.9)$$

$$F_z = F_c \sin \theta_h \quad (9.10)$$

where θ_h is the helical angle and $F'_c = F_c \cos \theta_h$.

To account for the helical angle, it is also assumed that the tool can be broken up into axial elements. Thus, the immersion angle $\varphi_j(t)$ shown in Figure 9.1 for each tooth and axial element must be determined to know if that tooth and axial element are engaged in the workpiece and thus contributing to the overall force of the system. The equation to find the immersion angle φ for each tooth, j, and axial element, k, is given in Equation (9.11)

$$\varphi_{jk}(t) = \left[\varphi_{ij} - \left(\frac{da}{2} + k\, da \right) \frac{\tan \theta_h}{R} + \frac{2\pi \Omega t}{60} \right] \quad (9.11)$$

It is assumed that the tool can be modeled as the lumped mass-spring-damper system seen in Figure 9.1. It is also assumed that because of the very high stiffness in the Z-direction, tool vibrations along the spindle axis are negligible. This results in two coupled equations of motion governing the X- and Y-direction motions of the tool as follows

$$m\ddot{x} + c_x \dot{x} + k_x x = F_x(t, \Delta x, \Delta y) \quad (9.12)$$

$$m\ddot{y} + c_y \dot{y} + k_y y = F_y(t, \Delta x, \Delta y) \quad (9.13)$$

9.3 Multivariable Micro-Milling Instability Control

The wavelet-based time-frequency nonlinear control law has been shown to address the fundamental characteristics inherent in dynamic instability, including bifurcation and chaos. Unlike modern control theories, which focus on eliminating time domain errors, the control

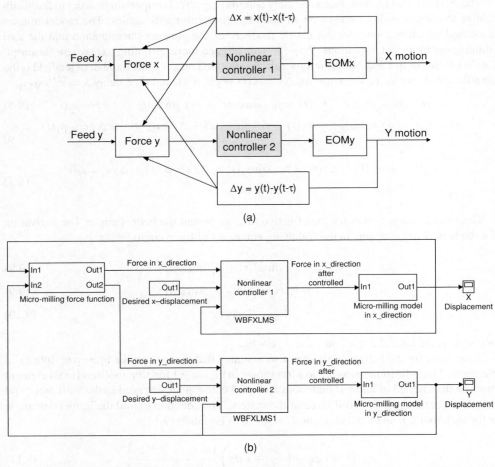

Figure 9.2 (a) Multivariable control configuration and (b) implementation algorithm

law restrains the deterioration of the time and frequency responses concurrently. The system response to be controlled is adjusted by the least-mean-square (LMS) adaptive filters to force the system to follow a target signal, which is the response of a stable state before dynamic deterioration. Because neither linearization nor a closed-form solution is required, all the genuine features of the nonlinear response are retained. The control scheme manipulates the corresponding discrete wavelet transform (DWT) coefficients of the system response to realize control in the joint time-frequency domain. The control theory is applied to the multidimensional micro-milling model described in Equations (9.1)–(9.13). By following the multivariable control architecture shown in Figure 9.2, the X- and Y-direction motions are determined using the force components that are defined by the feed and the instantaneous chip thickness in Equations (9.5)–(9.7). Figure 9.2(b) shows the implementation of the control configuration as a MATLAB® Simulink® algorithm. To achieve multivariable control, two independent nonlinear controllers are placed in front of the micro-milling model of each direction to mitigate the excitation force components. These two controllers operate in parallel and use different parameters.

The target signal is formulated using the truncated Fourier series of a desired micro-milling state of response. The desired state is defined by a stable vibration amplitude and a bounded frequency response containing the elementary modes that differentiate it from the instability states of bifurcation and chaos. When the controller is turned on, the system will be restored back to the desired stable state defined by the target signal. The construction of the target signal is briefly reviewed in the following, followed by an illustrative example in the next section. The Fourier series provides an alternate way of presenting a time signal by using harmonic functions of different frequencies. Suppose f is a T-periodic function defined within $[-T/2, T/2]$, then its Fourier series is

$$f(x) \sim \sum_{n=-\infty}^{\infty} c_n e^{2\pi i n/T} \tag{9.14}$$

and

$$c_n = \frac{1}{T} \int_{-T/2}^{T/2} f(x) e^{-2\pi i n/T} dx \tag{9.15}$$

Assume that $f(k)$ is the discrete form of $f(x)$ having N points within $[-T/2, T/2]$. The values of $f(k)$ outside the interval are assumed to be zeros. Then the Fourier expansion coefficient c_n can be represented as

$$c_n = \frac{1}{T} \frac{T}{N} \sum_{k=-\frac{N}{2}+1}^{\frac{N}{2}} f(k) e^{-\frac{2\pi i n k}{N}} = \frac{1}{N} F_n, \quad n = -\frac{N}{2}+1, \cdots \frac{N}{2} \tag{9.16}$$

where F_n is the Discrete Fourier Transform (DFT) of $f(k)$. A target signal can be reconstructed using Equation (9.14) by retaining only the frequency components in Equation (9.16) that represent the *fundamental modes* of a desired dynamic state that is physically stable.

9.3.1   Control Strategy

The model is simulated at a constant feed rate of 5 μm/tooth for different spindle speeds and axial depth-of-cuts (ADOC). The micro-dimensions used in [13] are adopted along with the pearlite material parameters found in [20]. The modal parameters for the tool are assumed to be equal in the X- and Y-direction, and are adopted from [21] since this research utilized a similar 500 μm micro-mill. These simulation parameters are summarized in Table 9.1.

An unstable response is observed for a spindle speed of 63 000 rpm and 100 μm ADOC. Under these cutting conditions, irregularity is observed in both the time and frequency responses using instantaneous frequency. The time response and IF of the X and Y motion are shown in Figure 9.3. A tool natural frequency at 4035 Hz and tooth passing frequency at 2100 Hz are observed. The tooth passing frequency is highly bifurcated as seen in the IF plots of Figure 9.3 that contain multiple frequency modes below the tooth passing frequency. When the spindle speed is reduced to 60 000 rpm and the ADOC is maintained at 100 μm, the

Table 9.1 Micro-milling parameters

Number of teeth	2
Tool diameter	500 μm
Tool nose radius, Re	2 μm
Rake angle	8°
Helical angle	30°
Tool natural frequency	4035 Hz
Stiffness	2142500 N/m
Damping ratio	0.016

tool response is one of stability (Fig. 9.4). The vibration response has improved with lower vibration amplitude and the IFs now show a stable dynamic response containing the 4035 Hz tool natural frequency and the 2000 Hz tooth passing frequency. The goal is to improve the dynamic stability of the process. Thus, it is desirable for the unstable response at 63 000 rpm

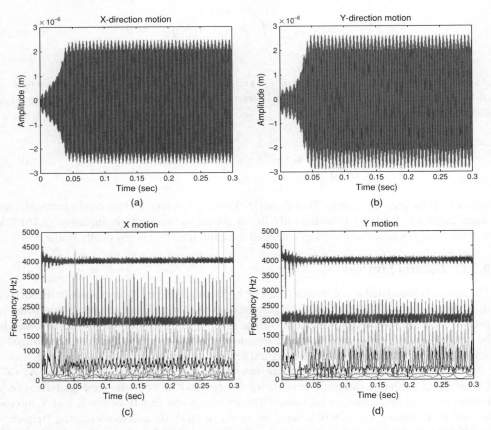

Figure 9.3 (a) Time response of x motion and (b) time response of y motion, and (c) IF of x motion and (d) IF of y motion at 63 000 rpm spindle speed and ADOC = 100 μm. Reproduced with permission from Steve C. Suh, Meng-Kun Liu, 2013, "Multi-Dimensional Time-Frequency Control of Micro-Milling Instability," Journal of Vibration and Control. Copyright 2013 Sage

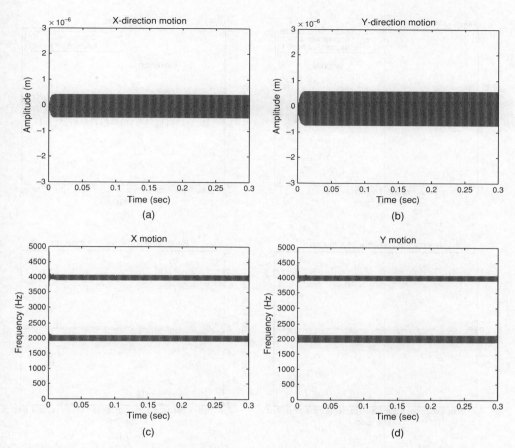

Figure 9.4   (a) Time response of x motion and (b) time response of y motion, and (c) IF of x motion and (d) IF of y motion at 60 000 rpm spindle speed and ADOC = 100 μm. Reproduced with permission from Steve C. Suh, Meng-Kun Liu, 2013, "Multi-Dimensional Time-Frequency Control of Micro-Milling Instability," Journal of Vibration and Control. Copyright 2013 Sage

and 100 μm axial DOC (Fig. 9.3) to better compare to the stable response at 60 000 rpm and 100 μm axial DOC (Fig. 9.4). The target signal for the controller is then developed based on the characteristics of the particular stable cutting. The target signal must contain the physically meaningful modes of the process as well as have acceptable vibration amplitudes. Then, the physically meaningful frequencies are retained while the undesirable frequency components are discarded.

For the 63 000 rpm case, the target signal contains vibration amplitudes similar to the 60 000 rpm stable response and consists of only the tool natural frequency and tooth passing frequency modes. The reconstructed signal and the reconstruction error of the signal can be seen in Figure 9.5. The reconstructed signal is fed into the controller as a reference signal. Figure 9.6 shows that unstable cutting at 63 000 rpm now has a controlled vibration and frequency response when the controller is turned on at 0.2 seconds. The vibration amplitude is reduced to a level similar to that of the stable cutting which will significantly improve

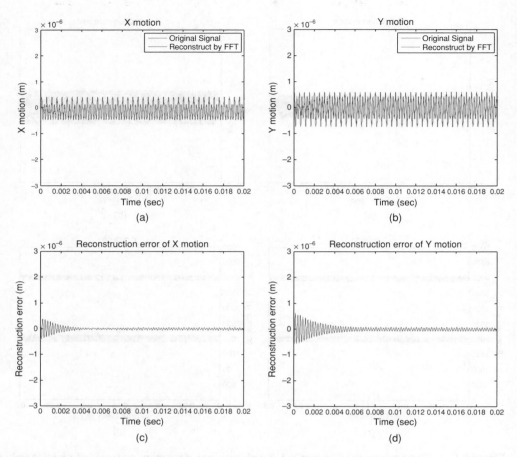

Figure 9.5 (a) Reconstructed target of x motion and (b) reconstructed target of y motion, and (c) reconstruction error of x motion and (d) reconstruction error of y motion. Reproduced with permission from Steve C. Suh, Meng-Kun Liu, 2013, "Multi-Dimensional Time-Frequency Control of Micro-Milling Instability," Journal of Vibration and Control. Copyright 2013 Sage

the workpiece tolerance and surface quality. The IF plot in Figure 9.6 also demonstrates an improved frequency response in which the individual modes are now bounded and range over a narrow bandwidth. The phase diagrams and Poincaré maps in Figure 9.7 indicate that the dynamic state of motion improves once the controller is initiated. The uncontrolled phase plot and Poincaré map indicate a fractal-like limit cycle while the controlled phase plot and Poincaré map demonstrate quasi-periodic motion with reduced amplitudes and a finite number of well-behaved frequency components.

9.4   Micro-Milling Instability Control

Control of the micro-milling model for different spindle speeds and ADOC are investigated with the assistance of phase portrait, Poincaré map, time response, IF, and cutting forces. Figure 9.8 shows the phase diagram and Poincaré map when the spindle speed is at 75 000 rpm with

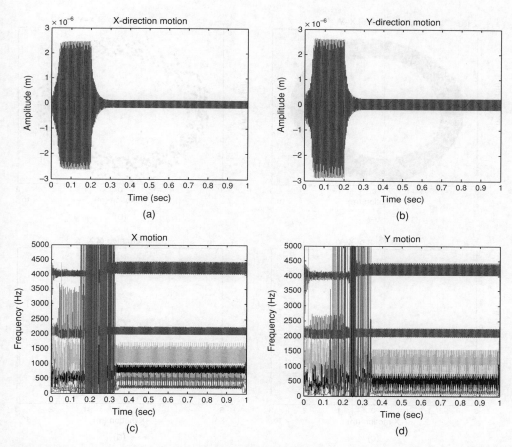

Figure 9.6 (a) Time response of x motion and (b) time response of y motion, and (c) IF of x motion and (d) IF of y motion when spindle speed = 63 000 rpm and ADOC = 100 μm. The controller is turned on at 0.2 seconds. Reproduced with permission from Steve C. Suh, Meng-Kun Liu, 2013, "Multi-Dimensional Time-Frequency Control of Micro-Milling Instability," Journal of Vibration and Control. Copyright 2013 Sage

an ADOC of 40 μm. Before the controller is activated, scattering on the Poincaré map in Figure 9.8(b) suggests that it is a broadband, chaotic response even though the time response is bounded. After being controlled, the phase plot in Figure 9.8(c) becomes a limit cycle with an amplitude that is four times smaller and the Poincaré map in Figure 9.8(d) shows a periodic motion of a finite number of commensurate frequencies. Figure 9.9 shows the time response and IF for controlling the milling process. The controller is turned on at 0.1 second and the target signal is designed using the response under the same spindle speed but a smaller ADOC of 30 μm. Before 0.1 seconds, the uncontrolled response has an irregular time response amplitude and broadband unstable frequency in both the x- and y-direction. Once the controller comes on-line at 0.1 seconds, the time response amplitude is reduced and the IF is restrained to a narrowband spectrum with a finite number of spectral components. The amplitude of cutting force in both directions is slightly reduced after being controlled as shown in Figure 9.10. The controller maintains the force amplitude around the stable cutting

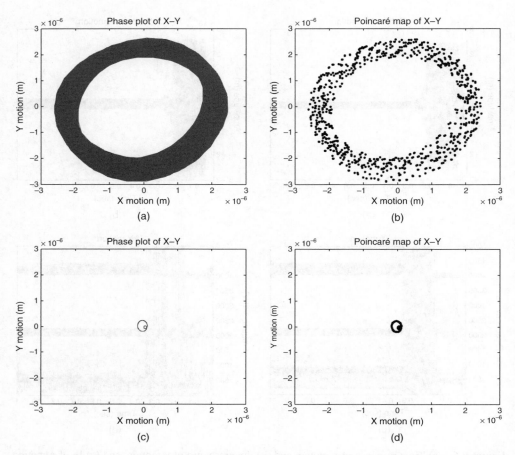

Figure 9.7 (a) Phase plot of x–y and (b) Poincaré map of x–y before being controlled, and (c) phase plot of x–y and (d) Poincaré map of x–y after being controlled when spindle speed = 63 000 rpm and ADOC = 100 μm. Reproduced with permission from Steve C. Suh, Meng-Kun Liu, 2013, "Multi-Dimensional Time-Frequency Control of Micro-Milling Instability," Journal of Vibration and Control. Copyright 2013 Sage

force limit for this particular chip load, effectively offsetting the negative impact of increased cutting forces due to dynamic instability.

When the spindle speed is increased to 90 000 rpm with 85 μm ADOC, its phase plot represents an unstable limit cycle in Figure 9.11(a) and the corresponding Poincaré map in Figure 9.11(b) shows a fractal structure. While not chaotic, it is of a broadband, varying spectrum and thus difficult to control. The target signal is composed from the same spindle speed with a 50 μm ADOC. When the controller is on, the response on the phase plot becomes one order-of-magnitude smaller and the Poincaré map becomes localized as shown in Figures 9.11(c) and 9.11(d), which means that the motion is now an attracting manifold having a frequency response whose bandwidth is significantly reduced. The time response and IF for controlling the milling process are shown in Figure 9.12. After the controller is applied at 0.1 second, the time response

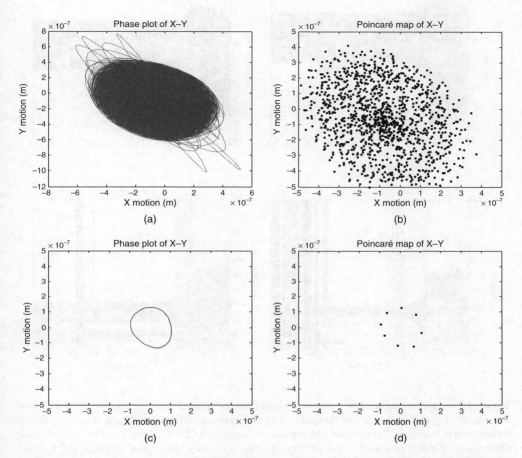

Figure 9.8   (a) Phase plot of x–y and (b) Poincaré map of x–y before being controlled, and (c) phase plot of x–y and (d) Poincaré map of x–y after being controlled when spindle speed = 75 000 rpm and ADOC = 40 μm. Reproduced with permission from Steve C. Suh, Meng-Kun Liu, 2013, "Multi-Dimensional Time-Frequency Control of Micro-Milling Instability," Journal of Vibration and Control. Copyright 2013 Sage

amplitude in both the X- and Y-direction is quickly decreased, thus helping to improve work-piece tolerance. The aberrational temporal oscillation of the instantaneous frequency is regularized and becomes a steady oscillation with limited bandwidth, as shown in Figure 9.12(c) and (d). The cutting force on both directions is effectively mitigated and restrained after being controlled (Fig. 9.13), thus improving the tool life of the process under these cutting conditions.

At a spindle speed of 180 000 rpm with ADOC of 50 μm, the time response amplitude is increased and becomes extremely irregular. For this case, the target signal is designed from the same spindle speed with a 30 μm ADOC. As all trajectories are seen being repelled from the manifold, the phase diagram and Poincaré map in Figure 9.14 show an unstable limit cycle with multiple frequencies before being controlled. After being controlled, the response is reduced to a (quasi) periodic motion with incommensurate frequencies demonstrating an improved dynamic state of motion. Both time response and instantaneous frequency are stabilized in

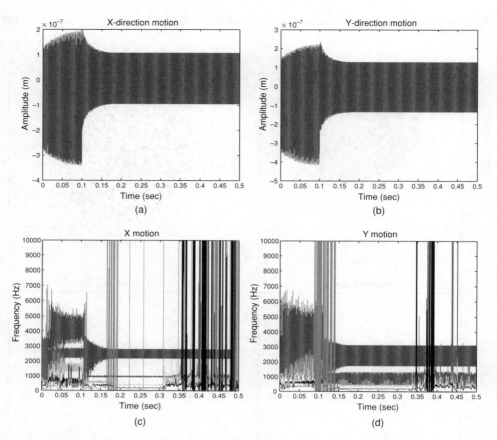

Figure 9.9 (a) Time response of x motion and (b) time response of y motion, and (c) IF of x motion and (d) IF of y motion when spindle speed = 75 000 rpm and ADOC = 40 μm. The controller is turned on at 0.2 seconds. Reproduced with permission from Steve C. Suh, Meng-Kun Liu, 2013, "Multi-Dimensional Time-Frequency Control of Micro-Milling Instability," Journal of Vibration and Control. Copyright 2013 Sage

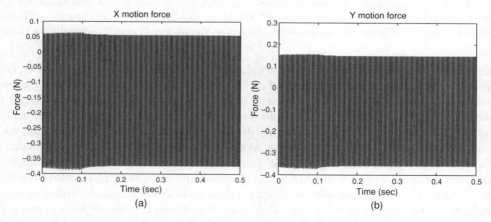

Figure 9.10 (a) Force in x-direction (b) force in y-direction when spindle speed = 75 000 rpm and ADOC = 40 μm. The controller is turned on at 0.1 seconds. Reproduced with permission from Steve C. Suh, Meng-Kun Liu, 2013, "Multi-Dimensional Time-Frequency Control of Micro-Milling Instability," Journal of Vibration and Control. Copyright 2013 Sage

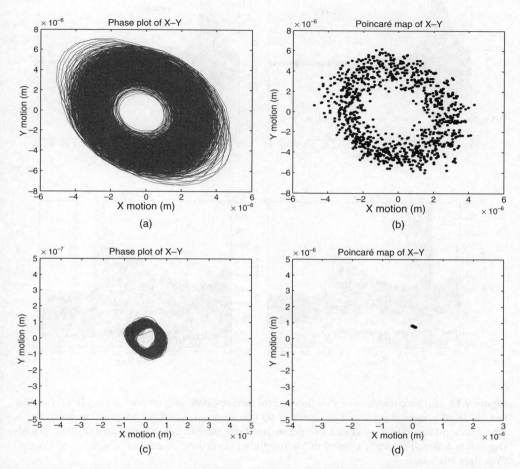

Figure 9.11   (a) Phase plot of x–y and (b) Poincaré map of x–y before being controlled, and (c) phase plot of x–y and (d) Poincaré map of x–y after being controlled at 90 000 rpm spindle speed and ADOC = 85 μm. Reproduced with permission from Steve C. Suh, Meng-Kun Liu, 2013, "Multi-Dimensional Time-Frequency Control of Micro-Milling Instability," Journal of Vibration and Control. Copyright 2013 Sage

Figure 9.15 after 0.1 seconds when the controller is brought on-line. Time response amplitude is reduced and aberrational oscillations in instantaneous frequency are erased when the controller is applied, resulting in narrowband frequency components. The cutting force is also reduced as shown in Figure 9.16 and maintains the stable cutting force limit.

When the spindle speed is 120 000 rpm with a 50 μm ADOC, resonance response is generated at the tooth passing frequency which is now coincident with the tool natural frequency. The phase plot and Poincaré map in Figures 9.17(a) and 9.17(b) show a periodic motion with a single frequency. After the controller is turned on, it indicates a reduction of amplitude in Figure 9.17(c). The time response and instantaneous frequency in Figure 9.18 show resonance amplitude and a signal frequency at 4000 Hz. The target signal is designed from the same spindle speed with 30 μm ADOC. The time response amplitude is reduced and its frequency

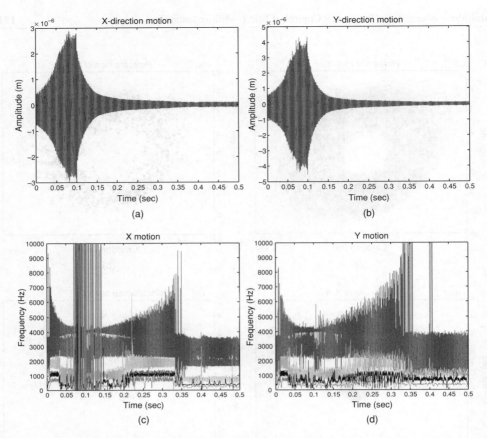

Figure 9.12 (a) time response of x motion and (b) time response of y motion, and (c) IF of x motion and (d) IF of y motion when spindle speed = 90 000 rpm and ADOC = 85 μm. The controller is turned on at 0.1 seconds. Reproduced with permission from Steve C. Suh, Meng-Kun Liu, 2013, "Multi-Dimensional Time-Frequency Control of Micro-Milling Instability," Journal of Vibration and Control. Copyright 2013 Sage

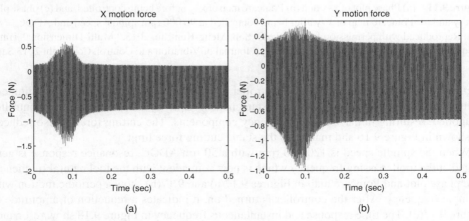

Figure 9.13 (a) Force in x-direction (b) force in y-direction when spindle speed = 90 000 rpm and ADOC = 85 μm. The controller is turned on at 0.1 seconds. Reproduced with permission from Steve C. Suh, Meng-Kun Liu, 2013, "Multi-Dimensional Time-Frequency Control of Micro-Milling Instability," Journal of Vibration and Control. Copyright 2013 Sage

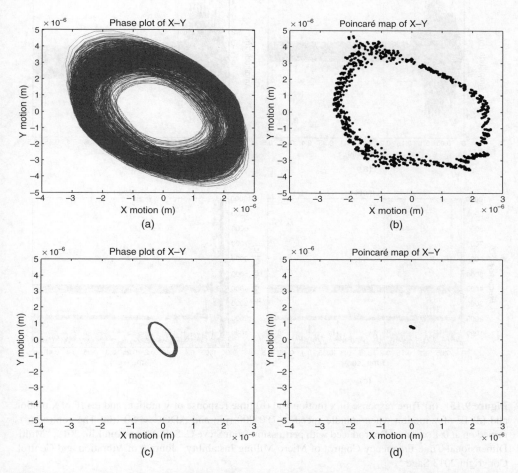

Figure 9.14 (a) Phase plot of x–y and (b) Poincaré map of x–y before being controlled, and (c) phase plot of x–y and (d) Poincaré map of x–y after being controlled when spindle speed = 180 000 rpm and ADOC = 50 μm. Reproduced with permission from Steve C. Suh, Meng-Kun Liu, 2013, "Multi-Dimensional Time-Frequency Control of Micro-Milling Instability," Journal of Vibration and Control. Copyright 2013 Sage

response remains the same after the controller is applied. The cutting force amplitude in both directions remains the same after being controlled (Figure 9.19). The reduction of the vibration amplitude is significant for improving product quality for high-speed cutting.

9.5 Summary

The micro-milling process is highly sensitive to dynamic instability which can result in premature tool breakage, increased wear rates, and poor workpiece quality. To achieve reasonable material removal rates while keeping the chip load at a minimum, high spindle speeds are

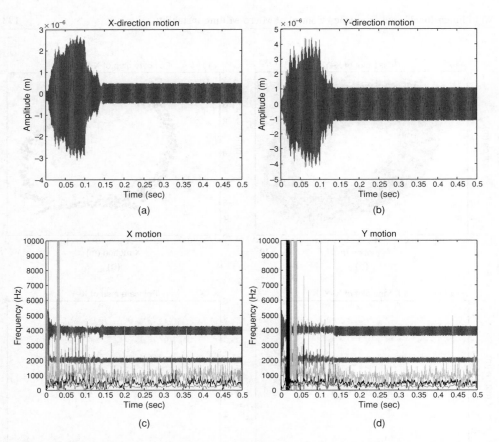

Figure 9.15 (a) Time response of x motion and (b) time response of y motion, and (c) IF of x motion and (d) IF of y motion when spindle speed = 180 000 rpm and ADOC = 50 μm. The controller is turned on at 0.1 seconds. Reproduced with permission from Steve C. Suh, Meng-Kun Liu, 2013, "Multi-Dimensional Time-Frequency Control of Micro-Milling Instability," Journal of Vibration and Control. Copyright 2013 Sage

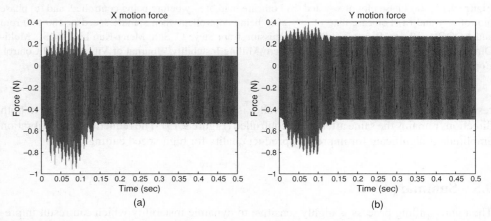

Figure 9.16 (a) Force in x-direction (b) force in y-direction when spindle speed = 180 000 rpm and ADOC = 50 μm. The controller is turned on at 0.1 seconds. Reproduced with permission from Steve C. Suh, Meng-Kun Liu, 2013, "Multi-Dimensional Time-Frequency Control of Micro-Milling Instability," Journal of Vibration and Control. Copyright 2013 Sage

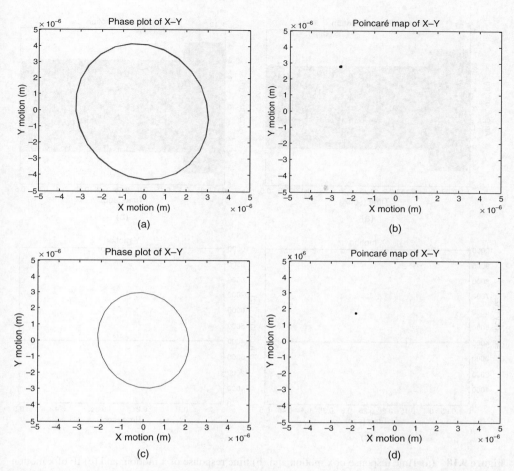

Figure 9.17 (a) Phase plot of x–y and (b) Poincaré map of x–y before being controlled, and (c) phase plot of x–y and (d) Poincaré map of x–y after being controlled when spindle speed = 120 000 rpm and ADOC = 50 μm. Reproduced with permission from Steve C. Suh, Meng-Kun Liu, 2013, "Multi-Dimensional Time-Frequency Control of Micro-Milling Instability," Journal of Vibration and Control. Copyright 2013 Sage

desired in micro-milling. This high frequency excitation of the system increases the effect of nonlinearity on the dynamic response, negatively impacting cutting performance by introducing increasingly broad frequency spectra as instability develops. This suggests that both the time and frequency domains should be considered to effectively control micro-milling, and a micro-milling model capable of capturing the high frequency signature of the process is required for testing the control algorithm.

The wavelet-based time-frequency controller was incorporated into the nonlinear multidimensional micro-milling model in order to control and improve the dynamic response under various spindle speeds and ADOC conditions, which resulted in an unstable dynamic state of

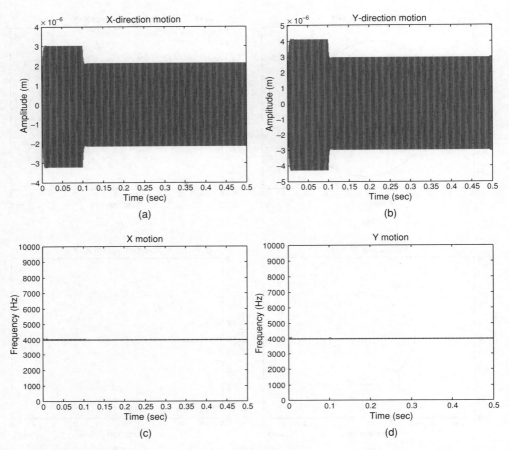

Figure 9.18 (a) Time response of x motion and (b) time response of x motion, and (c) IF of x motion and (d) IF of y motion when spindle speed = 120 000 rpm and ADOC = 50 μm. The controller is turned on at 0.1 seconds. Reproduced with permission from Steve C. Suh, Meng-Kun Liu, 2013, "Multi-Dimensional Time-Frequency Control of Micro-Milling Instability," Journal of Vibration and Control. Copyright 2013 Sage

motion. To control the process, two independent nonlinear controllers were placed in front of the model to regulate the cutting force excitation, and a target signal having all the physically meaningful frequency modes and acceptable amplitudes was utilized. The controller was applied to unstable cutting for spindle speed excitations ranging from 63 000 rpm to 180 000 rpm. For each case, the controller demonstrated the ability to reduce the vibration amplitude of the system, which is important for improving process efficiency and achieving and maintaining high precision cutting at a wide range of spindle speeds. The cutting forces were also observed to be properly mitigated and controlled to the stable cutting force values for that particular feed rate and ADOC. The controller prevented the negative effect of increasing cutting forces due to dynamic instability, thus simultaneously improving the life of the tool and negating immediate tool failure for unstable high speed excitation. The instantaneous frequency plots,

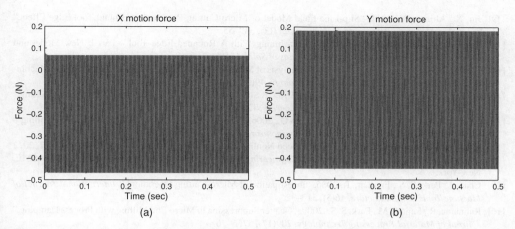

Figure 9.19 (a) Force in x-direction (b) force in y-direction when spindle speed = 120 000 rpm and ADOC = 50 μm. The controller is turned on at 0.1 seconds. Reproduced with permission from Steve C. Suh, Meng-Kun Liu, 2013, "Multi-Dimensional Time-Frequency Control of Micro-Milling Instability," *Journal of Vibration and Control*. Copyright 2013 Sage

phase portraits, and Poincaré maps illustrate the improved dynamic state of motion in the time-frequency domain after the controller is applied. This was observed by a reduction in the bandwidth of the frequency response, ultimately improving tool life and the wear rate. The application of the wavelet-based time-frequency controller to the highly nonlinear micro-milling process at high-speed excitation demonstrates the capability of mitigating the process in both the time and frequency domains with significantly improved tool performance and workpiece quality.

References

[1] Tansel, I., Rodriguez, M., Trujillo, E., Li, W., 1998, "Micro-End-Milling – I. Wear and Breakage," *International Journal of Machine Tools & Manufacture*, 38(12), 1419–36.

[2] Gandarias, S., Dimov, S., Pham, D. T., Ivanov, A., Popov, K., Lizarralde, R., Arrazola, P.J., 2006, "New Methods for Tool Failure Detection in Micromilling," *Proceedings of the Institution of Mechanical Engineers, Part B: Journal of Engineering Manufacture*, 220(2), 137–44.

[3] Byrne, G., Dornfeld, D., Denkena, B., 2003, "Advancing Cutting Technology," STC "C" Keynote, *CIRP Annals*, 52(2), 483–507.

[4] Basuray, P. K., Misra, B. K., Lal, G. K., 1977, "Transition From Ploughing to Cutting During Machining with Blunt Tools," *Wear*, 43(3), 341–49.

[5] Afazov, S. M., Ratchev, S. M., Segal, J., Popov, A. A., 2012, "Chatter Modeling in Micro-Milling by Considering Process Nonlinearities," *International Journal of Machine Tools & Manufacture*, 56(May), 28–38.

[6] Park, S., Rahnama, R., 2010, "Robust Chatter Stability in Micro-Milling Operations," *CIRP Annals – Manufacturing Technology*, 59(1), 391–94.

[7] Shi, Y., Mahr, F., Wagner, U., Uhlmann, E., 2012, "Chatter Frequencies of the Micromilling Processes: Influencing Factors and Online Detection via Piezoactuators," *International Journal of Machine Tools & Manufacture*, 56(May), 10–16.

[8] Jin, X., Altintas, Y., 2012, "Prediction of Micro-Milling Forces with Finite Element Method," *Journal of Materials Processing Technology*, 212(3), 542–52.

[9] Jin, X., Altintas, Y., 2011, "Slip-Line Field Model of Micro-Cutting Process with Round Tool Edge Effect," *Journal of Materials Processing Technology*, 211(3), 339–55.
[10] Fang, N., 2003, "Slip-Line Modeling of Machining With A Rounded Edge Tool – Part I: New Model and Theory," *Journal of the Mechanics and Physics of Solids*, 51(4), 715–42.
[11] Kim, J. D., Kim, D. S., 1995, "Theoretical Analysis of Micro-Cutting Characteristics in Ultra-Precision Machining," *Journal of Materials Processing Technology*, 49(3), 387–98.
[12] Waldorf, D. J., DeVor, R. E., Kapoor, S. G., 1998, "A Slip-Line Field for Ploughing During Orthogonal Cutting," *Journal of Manufacturing Science and Engineering*, 120(4), 693–99.
[13] Jun, M. B., Liu, X., DeVor, R. E., Kapoor, S. G., 2006, "Investigation of The Dynamics of Microend Milling – Part I: Model Development," *Journal of Manufacturing Science and Engineering*, 128(4), 893–900.
[14] Halfmann, E. B., Suh, C. S., 2012, "High Speed Nonlinear Micro Milling Dynamics," MSEC2012-7287, *2012 International Manufacturing Science and Engineering Conference (MSEC 2012)*, Notre Dame, IN, ASME, New York.
[15] Chae, J., Park, S. S., Freiheit, T., 2006, "Investigation of Micro-Cutting Operations," *International Journal of Machine Tools & Manufacture*, 46(3), 313–32.
[16] Rahnama, R., Sajjadi, M., Park, S. S., 2009, "Chatter Suppression in Micro End Milling with Process Damping," *Journal of Material Processing Technology*, 209(17), 5766–76.
[17] Fortgang, J., Singhose, W., de Juanes Márquez, J., Pérez, J., 2005, "Command Shaping for Micro-Mills and CNC Controllers," *American Control Conference*, Portland, OR, IEEE Press, 4531–36.
[18] Liu, P., Zhao, D., Zhang, L., Zhang, W., 2008, "Study of Cross-Coupling Control Approach Applied in Miniaturized NC Micro-Milling Machine Tool," *IEEE International Symposium on Knowledge Acquisition and Modeling Workshop*, IEEE Press, 888–91
[19] Aggogeri, F., Al-Bender, F., Brunner, B., Elsaid, M., Mazzola, M., Merlo, A., Ricciardi, D., de la O Rodriguez, M., Salvi, E., 2013, "Design of Piezo-based AVC System for Machine Tool Applications," *Mechanical Systems and Signal Processing*, 36(1), 53–65.
[20] Jun, M. B., DeVor, R. E., Kapoor, S. G., 2006, "Investigation of the Dynamics of Microend Milling – Part II: Model Validation and Interpretation," *Journal of Manufacturing Science and Engineering*, 128(4), 901–12.
[21] Malekian, M., Park, S., Jun, M., 2009, "Modeling of Dynamic Micro-Milling Cutting Forces," *International Journal of Machine Tools & Manufacture*, 49(7), 586–98.

10

Time-Frequency Control of Friction Induced Instability

As a severe contact process engaging the cutting tool with the chip, metal cutting is dynamically discontinuous involving friction. Friction-induced nonlinear vibrations and tool wear are among the factors that negatively impact workpiece quality and manufacturing yield [1]. In the chapter, friction-induced instability and discontinuity are studied. A flexible cantilever beam pressed against a rigid rotating disk will be explored in the sections that follow for studying self-excited friction-induced vibrations that are inherently unstable due to alternating friction conditions and decreasing dynamic friction characteristics. Knowledge established in the chapter has significant implications for the control of metal machining. Because no linearization or approximation scheme is followed, the genuine characteristics of the friction-disk model system, including stick-slip and inherent discontinuities, are fully disclosed without any distortion. We will see that the system dynamics are stable only within certain ranges of the relative velocity. With increasing relative velocity, the response loses its stability with diverging amplitude and broadening spectrum. The wavelet-based time-frequency control is subsequently applied to negate the chaotic vibrations at high relative velocity by adjusting the applied normal force. The controller design requires no closed-form solution or transfer function, hence allowing the underlying features of the discontinuous system to be fully established and properly controlled. Many prominent observations are made in the chapter, including that the inception of chaotic response at high relative velocity is effectively denied resulting in restoration of the system to a relatively stable state of limit-cycle.

10.1 Issues with Friction-Induced Vibration Control

Friction consists of two different states. In the state of "stick," the two contacting bodies are at rest and the static friction force acts against the start of the relative motion. Once there is relative motion, the state of "slip" characterized by a force-velocity curve having a negative slope at low relative velocities ensues. The friction reduces when the two contacting bodies

Control of Cutting Vibration and Machining Instability: A Time-Frequency Approach for Precision, Micro and Nano Machining, First Edition. C. Steve Suh and Meng-Kun Liu.
© 2013 John Wiley & Sons, Ltd. Published 2013 by John Wiley & Sons, Ltd.

start to move, but would again increase at high relative velocity. Friction-induced oscillations switch between these two states intermittently and display stick-slip motion [2]. The stick-slip motion can be considered the limit-cycle of a self-excited vibration system, where its stability is determined by the energy flowing into and dissipating from the system [3]. If the energy from the energy source flowing into the system is greater than the dissipated energy, the vibration amplitude increases. Otherwise the vibration amplitude decreases. A limit-cycle is formed when the energy input and dissipated energy during each period are in balance [4]. Few studies show that friction-induced oscillations could undergo subcritical bifurcation, and a slight intrusion into the unstable regime could result in large amplitude vibrations [5]. Friction-induced vibrations are the reason for brake squeal, excessive wear, fatigue, position inaccuracy, and often physical damage.

The mechanism for the generation of friction-induced vibration has been investigated experimentally [4]. It was shown that self-excited friction-induced vibrations could be caused by a decreasing friction characteristic, fluctuating normal force, geometrical effect, and nonconservative restoring force. When the dependence of the kinetic friction coefficient on relative velocity is of a negative slope, the steady sliding at equilibrium point becomes unstable and the instability leads to the generation of vibration [6]. Stick-slip vibrations generated by alternate friction models have also been researched numerically [7]. The shooting method was applied as a periodic solution finder in combination with the alternate friction model to locate periodic stick-slip solutions. Stick-slip motion and quasi-harmonic vibration were observed in phase portraits. Several methods for the control of friction induced vibrations were proposed in the literature. Time-delayed displacement feedback control force was applied to a mass-on-moving belt model in directions parallel and normal to the friction force [5]. It was shown to change the nature of the bifurcation from subcritical to supercritical, that is globally stable in the linearly stable regime. The friction induced vibration on a rotating disk was controlled by a time-delayed displacement feedback force [8]. It was able to reduce the vibration amplitude to nearly zero. A recursive time-delayed acceleration feedback control was applied to a mass-on-moving-belt model in [9]. The control signal was determined recursively by an infinite weighted sum of the acceleration of the vibrating system measured at regular time intervals in the past.

High frequency oscillations with small amplitudes were applied to a mass-on-moving-belt model parallel to the friction force in [2]. Adding high frequency harmonic excitation prevents self-excited oscillations from occurring. Instead of reducing self-excited oscillations to an absolute rest, the approach transformed oscillations into small amplitude vibrations at very high frequency. The dynamics of a Coulomb friction oscillator subjected to two harmonic excitations on a moving belt was investigated in [10]. The system dynamics was affected by the value of the frequency ratio and the amplitude of excitation. With a high frequency ratio and large amplitude, it was able to reduce chaotic motions to periodic motions. A fluctuating normal force consisting of a constant force and a superimposed sinusoidal force was applied to a moving belt model in [11] where the stick-slip motion was suppressed with reduced vibration amplitude.

A modulated normal load, in an on-off fashion depending upon the state of the system, was applied to a single DOF oscillator model on a moving belt in [12]. Derived from Lyapunov's second method to ensure dissipating energy, the control law was able to quench the unstable limit-cycle and transform it to a stable stick-slip limit-cycle. Active control law was used to vary the normal contact force in a joint by a piezoelectric actuator in [13]. The Lyapunov function-based control law was designed similar to a bang-bang controller to maximize energy dissipation instantaneously. The controller was shown to substantially reduce the vibration compared to joints with constant normal force. Active control techniques were adopted to improve the performance in eliminating the limit-cycle and the steady-state error caused by unknown friction and external disturbance [14]. An adaptive fuzzy inference system was employed to model the unknown friction dynamics and a proportional-derivative (PD) compensation controller was applied. A Lyapunov stability criterion was used to guarantee the convergence of the adaptive fuzzy model with PD controller.

However, all control methods reviewed previously have certain drawbacks. For time-delayed feedback control methods, the means to synthesize the control force and the amount of delayed time interval can only be determined heuristically. The same disadvantage is shared by applying high frequency oscillation and fluctuating normal force. They lack a systematic way to design the controller and predict performance. Because linear approximation is adopted to determine controller parameters for stability study, they all are unable to precisely determine the stability bounds of the friction-induced dynamics. The Lyapunov stability criterion is commonly applied to design the active control law. But the Lyapunov function candidate is difficult to come by for a complex system. Therefore, controller design based on the Lyapunov stability criterion is only applicable to simple systems or models that are significantly simplified through linear approximation. Nevertheless, linear approximation methods are unable to realize the genuine characteristics of any route-to-chaos process [15]. Because both time and frequency responses deteriorate simultaneously in route-to-chaos, it is crucial to design a nonlinear controller in the time and frequency domains concurrently. This is especially so for the highly nonlinear, intermittent friction-induced vibration system. A wavelet-based time-frequency control law was developed in [16] to restrain the concurrent time-frequency deterioration associated with the instability states of bifurcation and chaos. The controller has been shown in the previous chapter to be effective in denying milling chatters at high speed and restoring milling stability to a state of limit-cycle of extremely low tool vibration [17]. In the sections that follow, the finite element method along with a finite difference scheme is used to simulate the friction-induced vibration caused by pressing a flexible cantilever beam against a rigid rotating disk. Because no approximation method is adopted, the model preserves the underlying features of the friction induced route-to-chaos. The wavelet-based time-frequency control law from Chapter 7 will then be applied to the numerical model without resorting to a closed-form solution.

10.2   Continuous Rotating Disk Model

A simplified friction-induced vibration model from [8] is considered for the investigation. The model includes the discontinuity between the static and kinetic frictions and the dependence of the kinetic friction on the relative velocity. It consists of a cantilever beam with an end mass that is in frictional contact with a rigid rotating disk, as shown in Figure 10.1(a). The disk rotates at a constant angular velocity, ω_d. It is assumed to be rigid and displays no bending vibration or wobbling motion. A concentrated mass is attached to the end of the cantilever beam and a constant normal force F pushes the mass against the rigid disk, thus generating a friction force, F_r in the X–Y plane, that is also a function of the relative velocity, V_r.

The continuous rotating disk model is governed by

$$\frac{\partial^4 y}{\partial x^4} + \frac{\partial^2 y}{\partial \tau^2} = 0 \qquad\qquad (10.1)$$

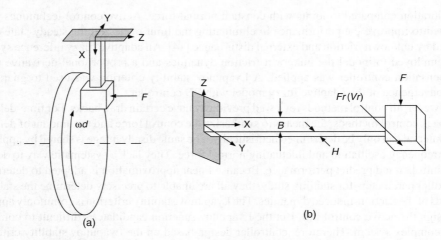

Figure 10.1 (a) Rotating disk model (b) flexible beam with an end mass

and subjected to the following boundary conditions

$$y(0, \tau) = 0 \quad , \quad \frac{\partial y}{\partial x}\bigg|_{x=0} = 0 \quad , \quad \frac{\partial^2 y}{\partial x^2}\bigg|_{x=1} = 0 \qquad (10.2)$$

and

$$\frac{\partial^2 y}{\partial \tau^2}\bigg|_{x=1} - r_m \frac{\partial^3 y}{\partial x^3}\bigg|_{x=1} = N_0 \mu(v_r) \qquad (10.3)$$

where the nondimensional variables are defined as

$$x = \frac{X}{L}, \quad y = \frac{Y}{L}, \quad \tau = \Omega t$$

$$\Omega = \sqrt{\frac{EI}{\rho A L^4}}, \quad r_m = \frac{\rho A L}{M}$$

$$N_0 = \frac{N_0^*}{M \Omega^2 L}, \quad v_d = \frac{V_d}{\Omega L}, \quad v_r = v_d - \frac{\partial y}{\partial \tau}\bigg|_{x=1}$$

in which

N_0^* is the externally applied normal load
L is the length of the beam
M is the mass of the end mass
E is the modulus of elasticity
ρ is the mass density of the beam material
$I = BH^3/12$ is the area moment of inertia of the beam cross-section
$A = BH$ is the cross-section area

v_d is the disk velocity
v_r is the normalized relative velocity between the disk and tip mass

   The coefficient of friction for the stick-slip motion follows the one formulated in [1] to prevent problems of slow convergence and numerical instability caused by the discontinuity attributable to zero relative velocity.

$$\mu(v_r) = \begin{cases} \dfrac{\mu_s v_r}{v_s}, & \text{for} \quad |v_r| < v_s \\[2mm] \mathrm{sgn}(v_r)\left[\mu_s - \dfrac{3}{2}(\mu_s - \mu_m)\left(\dfrac{|v_r| - v_s}{v_m - v_s} - \dfrac{1}{3}\left(\dfrac{|v_r| - v_s}{v_m - v_s}\right)^3\right)\right], & \text{for} \quad |v_r| \geq v_s \end{cases}$$

(10.4)

where

v_r is the nondimensional relative velocity between the mass and the disk
μ_s is the maximum coefficient of static friction
μ_m is the minimum coefficient of kinetic friction
v_m is the velocity corresponding to the minimum coefficient of kinetic friction μ_m
v_s is the velocity corresponding to the maximum coefficient of static friction μ_s
v_s is set to a small number at 0.0001. When the relative velocity $|v_r|$ is smaller than v_s, the model describes the state of sticking where the mass is at rest with respect to the moving disk. In contrast, when $|v_r| \geq v_s$, the state of slipping is described where the friction coefficient is a polynomial function. The *sign* function, $\mathrm{sgn}(v_r)$, can be replaced by $tan^{-1}(kv_r)$, $k \gg 1$. The friction function plotted using $\mu_s = 0.4$, $v_s = 10^{-4}$, $\mu_m = 0.25$, $v_m = 0.5$ is shown in Figure 10.2 as follows:

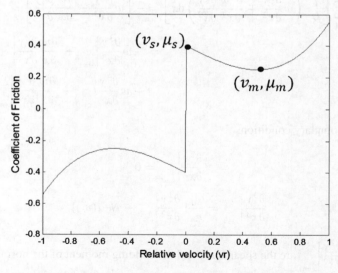

Figure 10.2   Friction coefficient as a function of relative velocity

To solve the boundary-value problem defined in Equations (10.1)–(10.3) numerically, Galerkin's method is applied to the normalized beam equation to derive the corresponding finite element formulation. The average weighted residual of Equation (10.1) is

$$I = \int_0^1 \left(\frac{\partial^2 y}{\partial \tau^2} + \frac{\partial^4 y}{\partial x^4} \right) w \, dx = 0 \tag{10.5}$$

where the length of the normalized beam is 1 and w is a weight function. The beam is discretized into a number of finite elements in Figure 10.3 below.

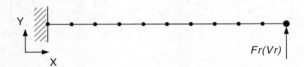

Figure 10.3   Discretization of the disk-brake model

With Ω^e defining the element domain and n the number of elements for the beam, Equation (10.5) becomes

$$I = \sum_{i=1}^{n} \left[\int_{\Omega^e} \left(w \frac{\partial^2 y}{\partial \tau^2} + \frac{\partial^2 w}{\partial x^2} \frac{\partial^2 y}{\partial x^2} \right) dx \right] + \left[w \frac{\partial^3 y}{\partial x^3} - \frac{\partial w}{\partial x} \frac{\partial^2 y}{\partial x^2} \right]_0^1 = 0 \tag{10.6}$$

Equation (10.6) can be rewritten as

$$
\begin{aligned}
\sum_{i=1}^{n} \left[\int_{\Omega^e} \left(w \frac{\partial^2 y}{\partial \tau^2} + \frac{\partial^2 w}{\partial x^2} \frac{\partial^2 y}{\partial x^2} \right) dx \right] &= - \left[w \frac{\partial^3 y}{\partial x^3} - \frac{\partial w}{\partial x} \frac{\partial^2 y}{\partial x^2} \right]_0^1 \\
&= -w \frac{\partial^3 y}{\partial x^3}\bigg|_{x=1} + \frac{\partial w}{\partial x} \frac{\partial^2 y}{\partial x^2}\bigg|_{x=1} \\
&\quad + w \frac{\partial^3 y}{\partial x^3}\bigg|_{x=0} - \frac{\partial w}{\partial x} \frac{\partial^2 y}{\partial x^2}\bigg|_{x=0}
\end{aligned}
\tag{10.7}
$$

subject to the boundary conditions

$$\frac{\partial^2 y}{\partial x^2}\bigg|_{x=1} = 0 \tag{10.8}$$

$$\frac{\partial^3 y}{\partial x^3}\bigg|_{x=1} = \frac{1}{r_m} [\frac{\partial^2 y}{\partial \tau^2}\bigg|_{x=1} - N_0 \mu(v_r)] \tag{10.9}$$

$\frac{\partial^3 y}{\partial x^3}\bigg|_{x=0}$ and $\frac{\partial^2 y}{\partial x^2}\bigg|_{x=0}$ are the shear force and the bending moment of the normalized beam at the fixed end, respectively. The Hermitian shape functions are used to interpolate the transverse deflection v in terms of nodal variables v_1, θ_1, v_2, and θ_2, shown using a two-node element in Figure 10.4 below.

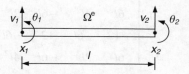

Figure 10.4 A two-node element

The transverse deflection v is therefore

$$v(x) = H_1(x)v_1 + H_2(x)\theta_1 + H_3(x)v_2 + H_4(x)\theta_2 \tag{10.10}$$

where

$$\mathbf{d} = [\, v_1 \quad \theta_1 \quad v_2 \quad \theta_2\,]^T$$

$$H_1(x) = 1 - \frac{3x^2}{l^2} + \frac{2x^3}{l^3}$$

$$H_2(x) = x - \frac{2x^2}{l} + \frac{x^3}{l^2}$$

$$H_3(x) = \frac{3x^2}{l^2} - \frac{2x^3}{l^3}$$

$$H_4(x) = -\frac{x^2}{l} + \frac{x^3}{l^2}$$

The dynamic equation of the two-node element is retrieved by applying the Hermitian shape function and Galerkin's method to the left-hand side of Equation (10.7)

$$[\mathbf{M_e}]\{\ddot{\mathbf{d}}_e\} + [\mathbf{K_e}]\{\mathbf{d}_e\} = \{\mathbf{F_e}(t)\} \tag{10.11}$$

Denoting $[\mathbf{H}] = \{\, H_1 \quad H_2 \quad H_3 \quad H_4\,\}$ and $[\mathbf{B}] = \{\, H_1'' \quad H_2'' \quad H_3'' \quad H_4''\,\}$, then

$$[\mathbf{M_e}] = \int_0^l [\mathbf{H}]^T [\mathbf{H}]dx$$

$$= \frac{r_m l}{420}
\begin{bmatrix}
156 & 22l & 54 & -13l \\
22l & 4l^2 & 13l & -3l^2 \\
54 & 13l & 156 & -22l \\
-13l & -3l^2 & -22l & 4l^2
\end{bmatrix} \tag{10.12}$$

$$[\mathbf{K_e}] = \int_0^l [\mathbf{B}]^T [\mathbf{B}] dx$$

$$= \frac{1}{l^3} \begin{bmatrix} 12 & 6l & -12 & 6l \\ 6l & 4l^2 & -6l & 2l^2 \\ -12 & -6l & 12 & -6l \\ 6l & 2l^2 & -6l & 4l^2 \end{bmatrix} \tag{10.13}$$

The internal force between the adjacent elements is cancelled out and only the external force is left to construct $\{\mathbf{F_e}(t)\}$. Only the boundary conditions need to be considered as the external force in this model. The global matrix equations (**M**, **K**, and **F**) for a dynamic beam analysis are assembled by summing up all corresponding element matrices and vectors (**M_e**, **K_e**, and **F_e**). The global equation of motion at time t becomes

$$[\mathbf{M}] \{\ddot{\mathbf{d}}\}^t + [\mathbf{K}] \{\mathbf{d}\}^t = \{\mathbf{F}\}^t \tag{10.14}$$

A finite difference scheme [18] is used to conduct transient analysis. Assume that the initial position and velocity are available. The initial acceleration can be calculated as

$$\{\ddot{\mathbf{d}}\}^0 = [\mathbf{M}]^{-1} \left\{ \{\mathbf{F}\}^0 - [\mathbf{K}] \{\mathbf{d}\}^0 \right\} \tag{10.15}$$

The velocity, displacement, and acceleration of Equation (10.14) at each time instance are approximated by calculating Equations (10.16)–(10.18) iteratively below,

$$\{\dot{\mathbf{d}}\}^{t+1} = \{\dot{\mathbf{d}}\}^t + \{\ddot{\mathbf{d}}\}^t \cdot \Delta t \tag{10.16}$$

$$\{\mathbf{d}\}^{t+1} = \{\mathbf{d}\}^t + \{\dot{\mathbf{d}}\}^{t+1} \cdot \Delta t \tag{10.17}$$

$$\{\ddot{\mathbf{d}}\}^{t+1} = [\mathbf{M}]^{-1} \left\{ \{\mathbf{F}\}^t - [\mathbf{K}] \{\mathbf{d}\}^{t+1} \right\} \tag{10.18}$$

10.3   Dynamics of Friction-Induced Vibration

To investigate the dynamics of friction-induced vibration, the following nondimensional parameters are considered: $r_m = 0.8$, $N_0 = 2$, $\mu_s = 0.4$, $\mu_m = 0.25$ and $v_m = 0.5$. The disk velocity, v_d, is the control parameter. Figure 10.2 is followed that correlates the friction coefficient with relative velocity. The dynamics of the tip mass corresponding to $v_d = 0.2$ is shown in Figure 10.5. After the response stabilizes, the velocity profile in Figure 10.5(a) conveys a clear stick-slip motion. The static friction dominates at certain periodic time periods when the tip mass moves at the same pace as the disk. Referring to Figure 10.2, it is seen that the friction coefficient is negatively proportional to the relative velocity in Figure 10.5(b). The self-excited oscillation forms a stable limit-cycle as shown in Figure 10.5(c). A static friction value at 0.8 is indicated in Figure 10.5(d) where the dynamic friction, whose value is smaller than the static friction, occurs intermittently throughout the time window. A dominant frequency of 0.1 Hz is observed in Figure 10.5(e). However, the instantaneous frequency in Figure 10.5(f)

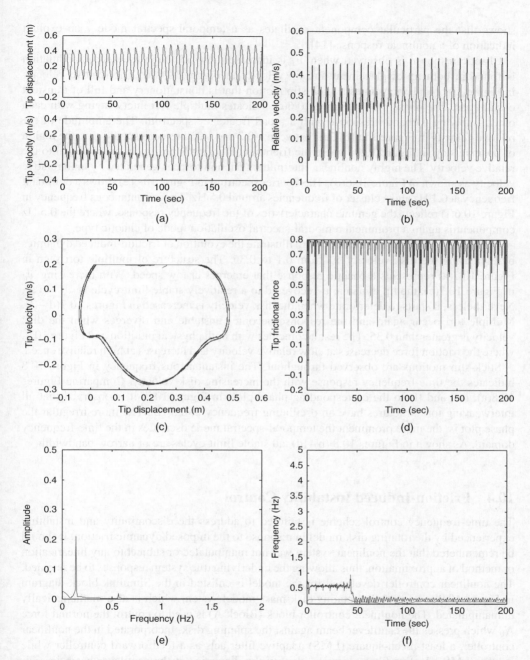

Figure 10.5 (a) Tip displacement and velocity (b) Relative velocity between tip mass and disk (c) Phase plot (d) Friction force in y-direction (e) Fourier spectrum (f) Instantaneous frequency of tip displacement when $v_d = 0.2$

shows that the particular component oscillates as a temporal-spectral mode – an explicit indication of a nonlinear response [14].

Chaotic response is observed when v_d is increased to 0.35. High frequency modulation is found in both tip displacement and tip velocity in Figure 10.6(a). The relative velocity in Figure 10.6(b) demonstrates a stick-slip motion that is nonstationary and full of transient oscillations. The phase plot in Figure 10.6(c) indicates multiple tori interweaving with each other, implying a chaotic response with a broad frequency spectrum. The amplitude of the relative velocity in Figure 10.6(b) falls between -0.2 and 1.0, thus corresponding to the region of positive slop in Figure 10.2, where the friction coefficient is positively proportional to the relative velocity. The highly nonlinear, intermittent friction force is generated by high relative velocity as shown in Figure 10.6(d). The Fourier spectrum in Figure 10.6(e) shows a dominant frequency at 0.1 Hz and a cluster of frequencies around 0.5 Hz. The instantaneous frequency in Figure 10.6(f) reflects the genuine characteristics of the frequency response, where the 0.5 Hz component is again a prominent temporal-spectral oscillation mode of chaotic type.

The phase portraits in Figure 10.7 are to illustrate the evolution of friction-induced dynamics by increasing the disk velocity v_d from 0.1 to 0.35. The structure of multiple tori seen in Figure 10.7(a) represents a chaotic response that emerges at low speed. With increasing v_d in Figure 10.7(c), the tip dynamic is recovered to a relatively stable limit-cycle at $v_d = 0.2$. Nevertheless, it starts to deteriorate when the disk velocity is increased in Figures 10.7(d)–(f). Multiple tori occur again and the response becomes unstable and diverges when the disk velocity is greater than 0.35. The result coincides with both physical intuition and Figure 10.2, where the friction force decreases at slow relative velocity and increases at high relative speed.

Stick-slip motions are observed throughout. The instantaneous frequency in Figure 10.8 indicates the time-frequency response with the increasing disk velocity. Comparing Figures 10.8(a), (e), and (f) to the corresponding phase plots in Figure 10.7, it is observed that all interweaving tori structures have an oscillating frequency at 0.5 Hz. The more irregular the phase plot is, the more prominent the temporal-spectral mode oscillates in the time-frequency domain. As shown in Figures 10.8 (b)–(d), all stable limit-cycles are of narrow bandwidth.

10.4   Friction-Induced Instability Control

The time-frequency control scheme is tailored to address the discontinuity and instability experienced by the rotating disk model in response to the imposed dynamic friction. It should be remembered that the nonlinear system was not manipulated or subject to any linearization or method of approximation, thus allowing the underlying true system responses to be retained. The nonlinear controller developed for the model is enlisted in the Simulink block diagram in Figure 10.9 to control the rotating disk-mass model system which is also mathematically unmanipulated. The nonlinear controller block (Block A) is used to control the normal force N_0 which presses the cantilever beam against the spinning disk. Incorporated in the nonlinear controller, a least-mean-square (LMS) adaptive filter acts as a feedforward controller while another LMS adaptive filter serves to identify the dynamics of the cantilever beam in real-time. Refer to Appendix A.1 for the accompanying MATLAB® modeling code. Both LMS adaptive filters manipulate the corresponding discrete wavelet transform (DWT) coefficients of the system response to realize control in the joint time-frequency domain. The force function block (Block B) calculates the tip friction force generated by multiplying the controlled normal

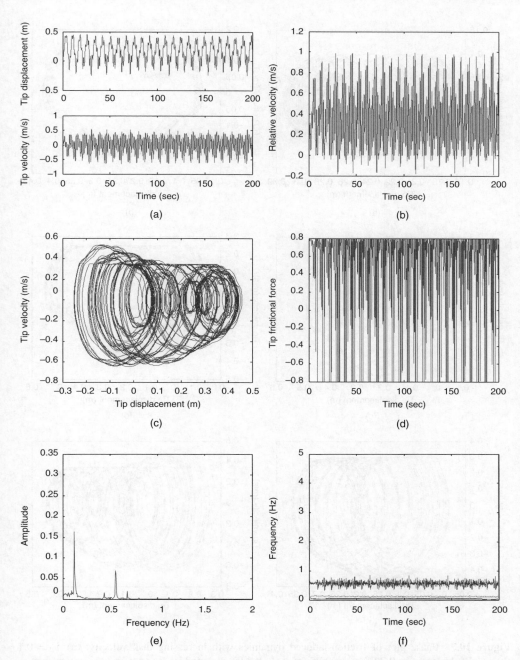

Figure 10.6   (a) Tip displacement and velocity (b) Relative velocity between tip mass and disk (c) Phase plot (d) Friction force in *y*-direction (e) Fourier spectrum (f) Instantaneous frequency of tip displacement when $v_d = 0.35$

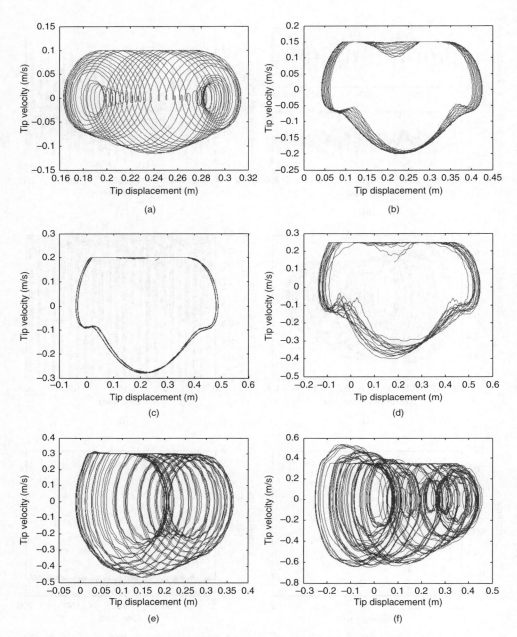

Figure 10.7 Phase plot of friction-induced dynamics with increasing disk velocity: (a) $v_d = 0.1$ (b) $v_d = 0.15$ (c) $v_d = 0.2$ (d) $v_d = 0.25$ (e) $v_d = 0.3$ (f) $v_d = 0.35$

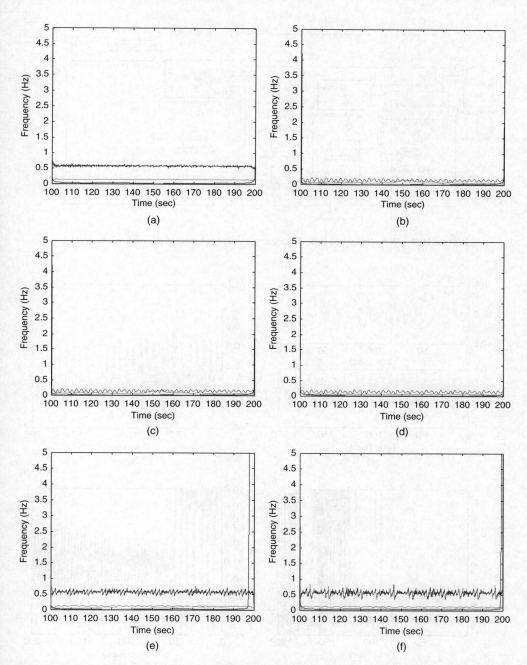

Figure 10.8 Instantaneous frequency of friction-induced dynamics with increasing disk velocity: (a) $v_d = 0.1$ (b) $v_d = 0.15$ (c) $v_d = 0.2$ (d) $v_d = 0.25$ (e) $v_d = 0.3$ (f) $v_d = 0.35$

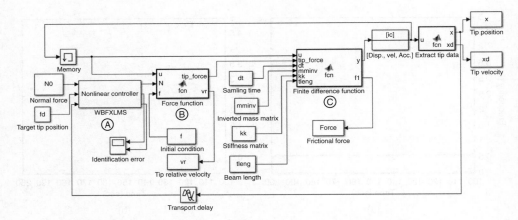

Figure 10.9   Wavelet-based time-frequency control scheme of the disk-brake model

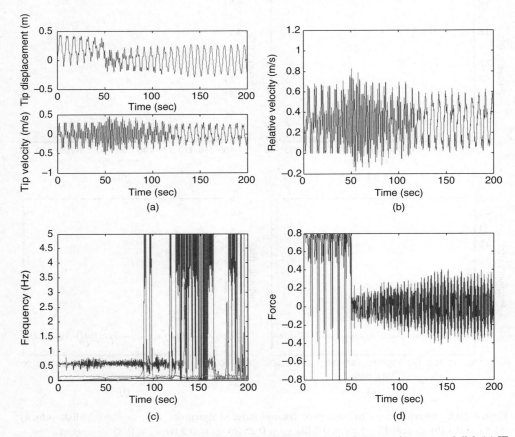

Figure 10.10   (a) Tip displacement and velocity (b) Relative velocity between tip mass and disk (c) IF (d) Friction force in y-direction when $v_d = 0.3$ and controller is turned on at 50 seconds

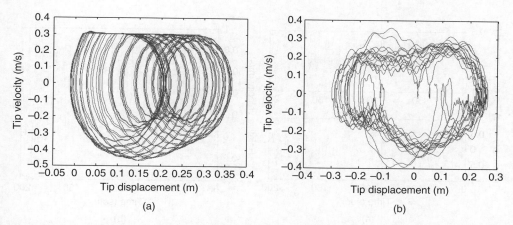

Figure 10.11 (a) Phase plot of tip movement before being controlled (b) after being controlled when $v_d = 0.3$

force with the coefficient of friction in Equation (10.4). The time-integration procedures found in Equations (10.15)–(10.18) are implemented in the finite difference function block, Block C, where the inverted mass and stiffness matrices are retrieved from the finite element model and the solution fields (i.e., displacement, velocity, and acceleration) of each node are calculated.

The control of two chaotic conditions when v_d is 0.3 and 0.35 is considered. Recall that chaotic vibrations were previously observed in Figure 10.7(e) where the disk velocity was 0.3. When the wavelet-based controller is turned on at 50 seconds, the oscillations of the tip displacement and velocity are rapidly abated to a more stable state of vibration, as seen in Figure 10.10(a). The applied controller intervenes in the system dynamics and effectively inhibits the emergence of stick-slip motions. The relative velocity in Figure 10.10(b) no longer lingers at zero, the state of static friction, and remains mostly in the state of dynamic friction. The nonlinear mode of temporal-spectral oscillations at 0.5 Hz in Figure 10.10(c) is eliminated after the controller is brought on-line, resulting in a narrowband response having only a finite number of distinct frequencies. The friction force in Figure 10.10(d) becomes smaller after being controlled, indicating that the stick-slip phenomenon becomes inconspicuous and opaque.

The phase plots of the tip motion before and after being controlled are compared in Figure 10.11. The trajectory after being controlled (Fig. 10.11(b)) shows a considerably more stable limit-cycle than Figure 10.11(a). It has a narrower bandwidth when the controller is applied, hence preventing the system dynamic from deteriorating further. Figures 10.12 and 10.13 show the controlled results in response to the same control strategy when v_d is increased to 0.35. Observations made of Figures 10.10 and 10.11 are readily applicable to Figures 10.12 and 10.13 where discontinuities such as stick-slip in the time domain and temporal-spectral oscillation in the time-frequency domain are effectively moderated. They all show that the wavelet-based time-frequency controller is highly effective in interfering in and interrupting the dynamics of the discontinuous system and, at the same time, restoring system chaotic response back to dynamic stability.

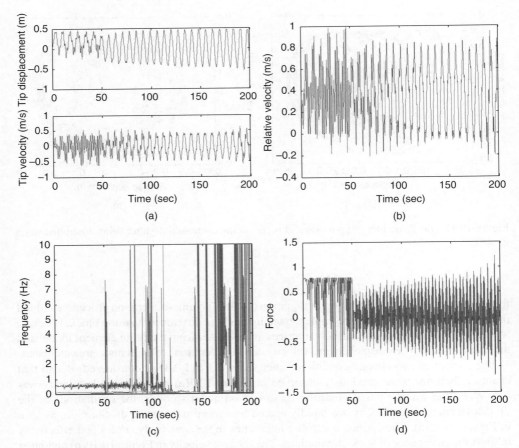

Figure 10.12 (a) Tip displacement and velocity (b) Relative velocity between tip mass and disk (c) Instantaneous frequency (d) Friction force in y-direction when $v_d = 0.35$ and the controller is turned on at 50 seconds

10.5 Summary

Unlike all the common studies on friction-induced dynamics, where either the linearization method or simplified model is adopted in order to conduct analytical calculations, the numerical investigation discussed in this chapter, employing the finite element method along with the finite difference time integration scheme, provided genuine, unaltered insight into the friction-induced vibrations. Stick-slip phenomena as a type of dynamic discontinuity were prominent and unduly accompanied by unstable limit-cycle and broadband chaotic response. Because no mathematical approximation or manipulation was attempted, responses modulated with high frequency components indicated rich nonlinearity. With the increase of the disk angular velocity, the modulated frequency became aberrational and broadband in the simultaneous time-frequency domain. The system lost its stability and the oscillation diverged when the disk angular velocity exceeded a certain critical value. The result coincides with the physical

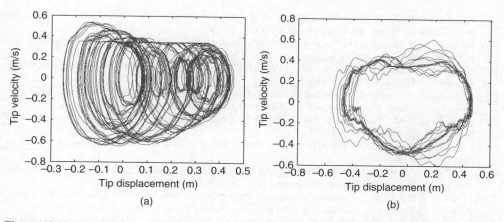

Figure 10.13 (a) Phase plot of the tip movement before being controlled (b) After being controlled when $v_d = 0.35$

interpretation of friction-induced vibration, which is highly nonlinear due to the discontinuity of the two moving surfaces of contact. When the friction coefficient was negatively proportional to the relative velocity, the self-excited vibration generated a limit-cycle. The system became unstable at high relative velocity when the friction coefficient was positively proportional to the relative velocity. The deployed wavelet-based time-frequency controller was effective in controlling the disk-brake model. It was applied to the finite element model directly without resorting to linearization or closed-form solution, thus allowing the intrinsic features of the friction-induced dynamics to be captured and controlled. The controller adequately inhibited the inauguration of chaotic response at high relative velocity and returned the system to a relatively stable limit-cycle. Because the controller was designed in the time-frequency domain, it suppressed the diverging vibration amplitude and increasing bandwidth concurrently. The result indicated that time-frequency control is a viable solution to mitigating friction-induced instabilities including route-to-chaos and friction-induced dynamic discontinuity.

References

[1] Childs, T.H.C., 2006, "Friction Modeling in Metal Cutting," *Wear*, 260(3), 310–18.
[2] Thomsen, J.J., 1999, "Using Fast Vibration to Quench Friction-Induced Oscillations," *Journal of Sound and Vibration*, 228(5), 1079–1102.
[3] K. Popp, M. Rudolph, 2004, "Vibration Control to Avoid Stick-Slip Motion," *Journal of Vibration and Control*, 10(11), 1585–1600.
[4] M. Kröger, M. Neubauer, K. Popp, 2008, "Experimental Investigation On The Avoidance of Self-Excited Vibrations," *Philosophical Transaction of the Royal Society, A* 366(1866), 785–810.
[5] A. Saha, P. Wahi, 2011, "Delayed Feedback for Controlling The Nature of Bifurcations in Friction-Induced Vibrations," *Journal of Sound and Vibration*, 330(25), 6070–87.
[6] K. Nakano, S. Maegawa, 2009, "Safety-Design Criteria of Sliding Systems for Preventing Friction-Induced Vibration," *Journal of Sound and Vibration*, 324(3), 539–55.
[7] R.I. Leine, D.H. Van Campen, A. De Kraker, L. Van Den Steen, 1998, "Stick-Slip Vibrations Induced By Alternate Friction Models, *Nonlinear Dynamics*, 16(1), 41–54.

[8] A. Saha, S.S. Pandy, B. Bhattacharya, P. Wahi, 2011, "Analysis and Control of Friction-Induced Oscillations in A Continuous System, *Journal of Vibration and Control*, 18(3), 467–80.

[9] S. Chatterjee, P. Mahata, 2009, "Controlling Friction-Induced Instability by Recursive Time-Delayed, *Journal of Sound and Vibration*, 328(2), 9–28.

[10] G. Cheng, J.W. Zu, 2003, "Dynamics of A Dry Friction Oscillator Under Two-Frequency Excitation, *Journal of Sound and Vibration*, 275(3), 591–603.

[11] M. Neubauer, C. Neuber, K. Popp, 2005, "Control of Stick-Slip Vibrations," *IUTAM Symposium on Vibration Control of Nonlinear Mechanisms and Structures*, Springer, Munich, 223–32.

[12] S. Chatterjee, 2007, "Nonlinear Control of Friction-Induced Self-Excited Vibration," *International Journal of Non-Linear Mechanics*, 42(3), 459–69.

[13] L. Gaul, R. Nitsche, 2000, "Friction Control for Vibration Suppression," *Mechanical System and Signal Processing*, 14(2), 139–50.

[14] Y.F. Wang, D.H. Wang, T.Y. Chai, 2011, "Active Control of Friction-Induced Self-Excited Vibration Using Adapting Fuzzy Systems, *Journal of Sound and Vibration*, 330(17), 4201–10.

[15] M.K. Liu, C.S. Suh, 2012, "Temporal and Spectral Responses of A Softening Duffing Oscillator Undergoing Route-To-Chaos," *Communications in Nonlinear Science and Numerical Simulations*, 17(12), 5217–28.

[16] M.K. Liu, C.S. Suh, 2012, "Simultaneous Time-FrequencyC of Bifurcation and Chaos," *Communications in Nonlinear Science and Numerical Simulations*, 17(6), 2539–50.

[17] M.K. Liu, C.S. Suh, 2012, "On Controlling Milling Instability and Chatter At High Speed," *Journal of Applied Nonlinear Dynamics*, 1(1), 59–72.

[18] Y.W. Kwon, D. Salinas, M.J. Neibert, 1994, "Thermally Induced Stresses in A Trilayered System, *Journal of Thermal Stresses*, 17(3), 489–506.

11

Synchronization of Chaos in Simultaneous Time-Frequency Domain

Chatter as a detrimental mode in metal cutting is a strong nonlinear function of the complex chip-forming process that involves thermoplasticity, machine tool stiffness, regenerative effect, and dynamic discontinuity – such as the intermittent engagement of the tool with the workpiece. We concluded in Chapter 10 that velocity-dependent friction and the resulted stick-slip motions result in prominent chaotic motion. These types of dynamic discontinuities are particularly pronounced in frictional chatter [1]. Suppressing self-sustained and chaotic machining chatter usually takes inspiration from chaos control. It is possible to control chatter and cutting instability by the approach commonly known as synchronization of chaos. As is implied by [2], synchronization between a reference-driven model and the real cutting process (the driving) can be a viable solution to rejecting chatter through maintaining a limit-cycle type of stability. The notion of exerting concurrent control in the joint time-frequency domain is applied in the present chapter to formulate a chaos synchronization scheme that would have broad implications for chatter control. The chaos synchronization scheme requires no linearization or heuristic trial-and-error for the design of the nonlinear controller. Without *a priori* knowledge of the driven system parameters, synchronization is invariably achieved regardless of the initial and forcing conditions the response system is subjected to. In addition, driving and driven trajectories are seen to be robustly synchronized with negligible errors in spite of the infliction of high frequency noise. We begin the chapter with a review of chaos synchronization.

11.1 Synchronization of Chaos

Chaos synchronization can be categorized into complete, practical, partial, and almost synchronization [3, 4]. The simplest synchronization method is complete replacement, which

Control of Cutting Vibration and Machining Instability: A Time-Frequency Approach for Precision, Micro and Nano Machining, First Edition. C. Steve Suh and Meng-Kun Liu.
© 2013 John Wiley & Sons, Ltd. Published 2013 by John Wiley & Sons, Ltd.

substitutes the variable in the response system with the corresponding variable being passed from the drive system [3]. The stable synchronization can also be achieved by replacing the variable only in certain locations, called partial replacement. The drive and response systems can also be coupled by adding a damping term that consists of a difference between the drive and the response variables. However, most studies have the configuration of the synchronization scheme determined through trial by error.

Lyapunov stability theory is often adopted when formulating chaos synchronization methods. Proper rule of update for unknown parameters and control law for compensating the external excitation are designed to make the Lyapunov function candidate compatible with the stability requirement [5]. Synchronization can be robustly achieved for identical or dissimilar chaotic systems without the calculation of the conditional Lyapunov exponents. An adaptive back-stepping control law is applied to ensure that the error between the drive and response system is asymptotically stable [6, 7]. It derives the Lyapunov function candidate in sequence for each variable, progressively stepping back from the overall system and securing the stability for each variable. To identify the chaotic system simultaneously, adaptive control law is derived from Lyapunov theory to define the convergence and stability of the error dynamic equation [8, 9, 10]. It is able to accommodate unknown parameters and system uncertainty. An adaptive sliding mode controller is used to synchronize the nonautonomous system with a sinusoid driving term [11, 12]. Designed to make the derivative of the Lyapunov function negative, the switching control law is formulated from Lyapunov theory to guarantee the asymptotical stability and convergence on the sliding surface of the error state space equation. The continuous input thus obtained can withstand uncertainties and disturbances. This is a quality common to all synchronization methods. Nevertheless, control laws thus formulated are neither intuitive nor lend themselves to sound physical interpretation. In addition, they require that the system structure and parameters be explicitly known. This requirement is even more mandatory for nonautonomous and nonstationary systems, so that the control law can be properly designed to cancel out the external forcing term. The Lyapunov function needs to be found based on heuristic methods and can only be applied to relatively simple systems. Thus, methods based on the Lyapunov stability theory are not viable for complex, nonstationary systems of unknown or unspecified nature.

Linear control theories are also applied to synchronization of chaos. Two different chaotic systems are synchronized by assigning proper control law to make the error dynamic equation linear [13, 14]. Linear stability theory, such as calculating the eigenvalues of the state matrix, is used to define stability. Linear feedback control is applied to synchronize two identical chaotic systems [15]. The system is linearized by Jacobin and the Routh–Hurwitz stability criterion is followed to identify the feedback gain to suppress chaos to unstable equilibria. Two linearly coupled chaotic systems are synchronized by assigning the proper parameters to satisfy the Routh–Hurwitz stability criterion [16]. The synchronization method of the particular construct either designs a linear error dynamic equation by choosing a specific drive-response chaotic system pair or adopts Jacobin to the state matrix. However, a linear error dynamic equation can only be acquired for a limited drive-response pair and uncertainties of the system are not allowed. As such it is basically a trial-and-error method with limited applicability. Jacobin is in fact a linearization method that can only be applied to the adjacent area of the pre-determined fixed point. Its sensitivity to disturbances renders it suitable only for stationary systems. For nonstationary systems, once the trajectory deviates away from the fixed point, the stability is no longer valid.

Synchronization of delayed differential equations is also reported, including identical dissipative chaotic systems with nonlinear time-delayed feedback that are unidirectionally coupled and synchronized [17]. The synchronization threshold of the coupled time-delay chaotic system is analytically estimated by calculating the Lyapunov exponents [18]. A time-delayed feedback term is added into the control law to synchronize two identical Lur systems by using the Lyapunov stability theory to ensure stability [19, 20, 21]. The delay time is often heuristically decided, however. If more than one variable needs to be passed to the response system, synchronous substitution is used to define a new variable as a function of multiple variables [22]. This new variable is transmitted to the response system and recovered by inverting the transformation. The synchronization of a nonautonomous chaotic system in [23] specifies an identical sinusoidal forcing term of different phases to both the drive and response systems. A strobe signal is used to form a feedback loop and modulate the frequency of the function generator to provide the sinusoidal excitation for the response circuit. In general, chaos synchronization methods of late, though functional, have limited applications only for the well-defined situations.

11.2 Dynamics of a Nonautonomous Chaotic System

The wavelet-based time-frequency control formulated in Chapter 7 is applied in the following sections to synchronize a nonautonomous chaotic circuit [23] subject to different external excitations. The circuit, which is a periodically forced chaotic system preferred for its superior insensitivity to noise than the autonomous systems, is governed by the system of three equations below

$$\frac{dx}{dt} = \beta (y - z) \tag{11.1}$$

$$\frac{dy}{dt} = \beta \left(-\Gamma_y y - g(x) + \alpha \cos(\omega t) \right) \tag{11.2}$$

$$\frac{dz}{dt} = \beta (f(x) - \Gamma_z z) \tag{11.3}$$

where

$$g(x) = -3.8 + (1/2)(|x + 26| + |x - 2.6| + |x + 1.2| + |x - 1.2|) \tag{11.4}$$

$$f(x) = \frac{1}{2}x + |x - 1| - |x + 1| \tag{11.5}$$

$g(x)$ and $f(x)$ are piecewise linear functions based on a diode function generator. Parameters $\alpha = 0.2$, $\Gamma_y = 0.2$, $\Gamma_z = 0.1$, $\beta = 10^4$, and $\omega = 2\pi f_d$ with the linear forcing frequency $f_d = 769$ Hz [23]. The circuit is numerically time-integrated using a 0.0001 sec time step subject to the $[x \ y \ z] = [0\ 0\ 0]$ initial conditions. Figure 11.1 shows the phase diagram and Poincaré map of the chaotic responses in the x–y, x–z, and y–z planes.

The phase portraits in Figures 11.1(a)–(c) have limit-cycle-like trajectories overlapping each other, representing a nonstationary frequency oscillation with a limited spectral bandwidth; while the Poincaré maps in Figures 11.1(d)–(f) demonstrate fractal structures in each

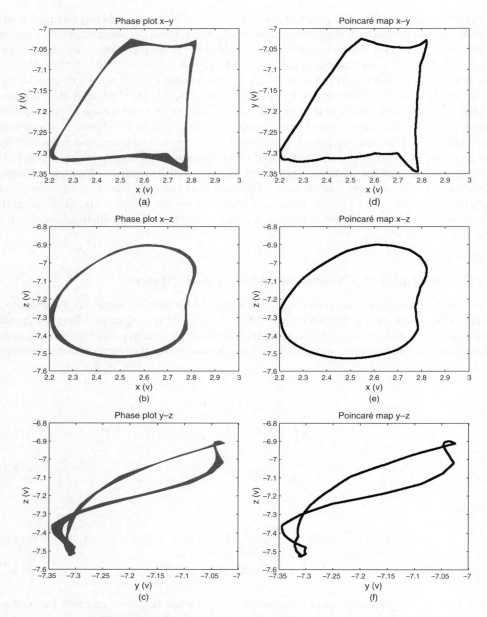

Figure 11.1   Phase diagram (left) and Poincaré map (right)

dimension. To resolve the hidden nonlinearity indicated in the figure for the nonautonomous chaotic circuit, instantaneous frequency (IF) along with marginal spectrum is applied in the following.

The IF of the x motion in Figure 11.2(a) shows prominent temporal oscillations of the first frequency mode along with other aberrational irregular modes. The corresponding marginal

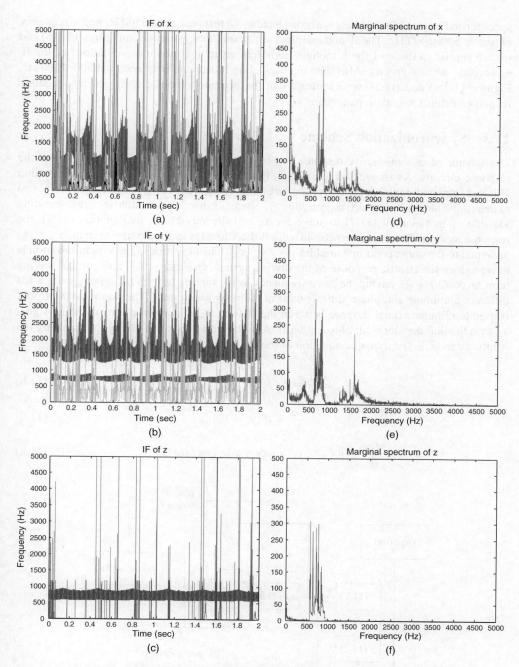

Figure 11.2 Instantaneous frequency (IF) (left) and marginal spectrum (right)

spectrum in Figure 11.2(d) indicates a broad bandwidth response up to 2500 Hz and a frequency cluster at 500–1000 Hz. The IF and marginal spectrum of the y motion in Figures 11.2(b) and (e) are similar to those of the x motion, both demonstrating temporal oscillations of the IF and broad marginal spectra. Also showing oscillating broadband frequencies at 500–1000 Hz, Figures 11.2(c) and (f) likewise indicate that the nonautonomous circuit generates chaotic responses subject to certain parameters.

11.3 Synchronization Scheme

The scheme of cascaded drive-response in [24] is adopted to synchronize the drive and the response circuits. As shown in Figure 11.3, the variable x of the drive system in Equation (11.1), completely replaces the corresponding variable x_1 in the subsystem in Equation (11.6) to determine the secondary driving variable z_1. Then z_1 is used to substitute the corresponding variable z_2 in Equation (11.7) to solve for x_2 and the variable y_2 in Equation (11.8), the response system. Thus when synchronized, all the variables in the driving system (x, y, z) are equal to the corresponding variables (x_2, y_2, z_2). The objective of the synchronization is to reproduce the chaotic response of the driving system. In Figure 11.3, a sinusoidal forcing term, $\alpha_r \cos(\omega t + \theta)$, having the same frequency as the forcing term in the driving system but different amplitude and phase shift, is used in the response system. The nonlinear controller is used to eliminate the difference between the chaotic signal x and the corresponding signal x_2 by adjusting the sinusoidal forcing term. A high frequency noise during the transmission is infused as $d(t)$. The dynamic equation of the response system is as follows,

$$\frac{dz_1}{dt} = \beta \left(f(x_1) - \Gamma_z z_1 \right) \tag{11.6}$$

$$\frac{dx_2}{dt} = \beta \left(y_2 - z_2 \right) \tag{11.7}$$

$$\frac{dy_2}{dt} = \beta \left(-\Gamma_y y_2 - g(x_2) + \alpha_r \cos(\omega_r t + \theta) \right) \tag{11.8}$$

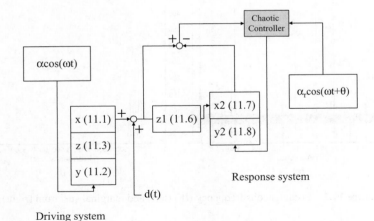

Figure 11.3   Synchronization scheme

11.4   Chaos Control

11.4.1   Scenario I

The drive and response systems are subject to different initial conditions and driving forces of dissimilar amplitudes and phases. The chaotic circuit in Equations (11.1)–(11.3) is the drive system and the variable x is transmitted to the response system defined in Equations (11.6)–(11.8). The initial conditions specified for the response system are $[x, y, z] = [2, 2, 2]$ and the driving term is set as $\alpha_r \cos(2\pi \omega_d + \theta)$, with $\alpha_r = 0.4$, $\omega_d = 769$, and $\theta = \pi/2$. To observe the difference between the drive and the response systems, their phase portraits are drawn on the same scale and compared in Figure 11.4. Their output trajectories locate at the same basin in the state space. It is shown that the output of the drive and the response systems share the same traits but are of different magnitudes, indicating that they are unsynchronized and of different chaotic responses.

A Daubechies-3 (db3) wavelet with a 1st decomposition level is deployed to construct the transformation matrix T. (See Chapter 7.) The control scheme for the synchronization problem at hand is given in Figure 11.6 as a block diagram where the wavelet-based time-frequency controller is denoted as WBFXLMS. Refer to Appendix A.2 for the accompanying MATLAB® modeling code. The left column in Figure 11.5 shows the output of the response system. When

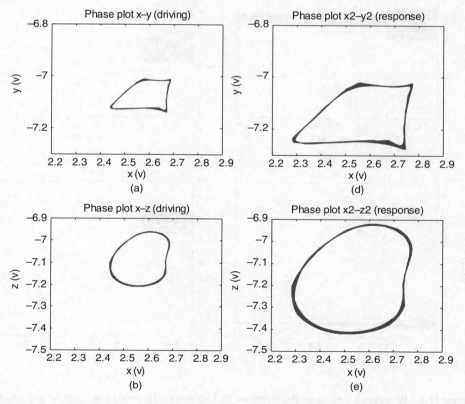

Figure 11.4   Phase diagrams of the output of the drive (left) and response (right) systems (*Continued*)

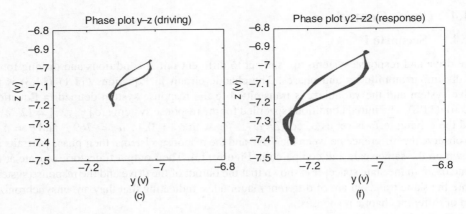

Figure 11.4   (*Continued*)

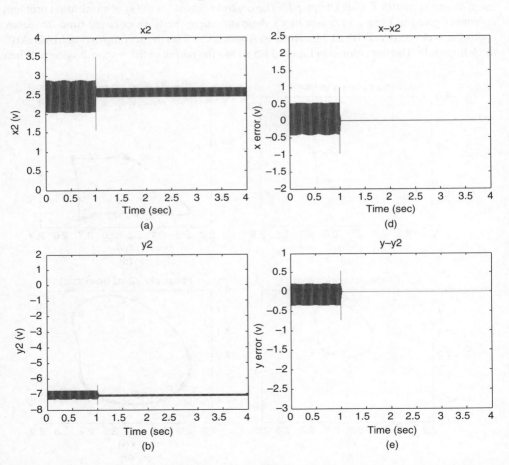

Figure 11.5   Response when controller is applied at 1 second (left); error between drive and response signals (right) (*Continued*)

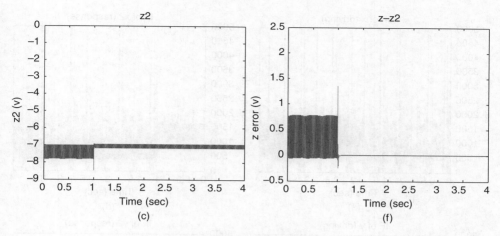

Figure 11.5 (*Continued*)

the controller is turned on at the 1 second mark, the output of the response system is converged to match the output of the drive system after a short transient. The three plots on the right indicate that the error between the drive and the response systems is reduced to almost zero after the controller is brought on-line. To observe the influence on the frequency response by the controller, the instantaneous frequency (IF) of the controlled response signal in the right column of Figures 11.7 is compared with the driving signal, which is in the left column of the same figure. For clarity, only the first one or two frequency modes are illustrated. It is seen that the IF of the response signal is restored to follow the driving signal with great fidelity after the controller is activated at 1 second. That the two oscillators share the same IF characteristics is a strong indication that the wavelet-based controller is highly effective in synchronizing the drive-response system with conspicuous correspondence and accuracy in both the time and frequency domains.

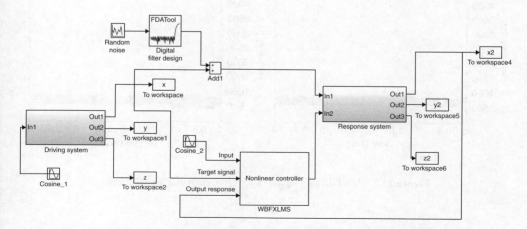

Figure 11.6   Control scheme with wavelet-based time-frequency (WBFXLMS) controller

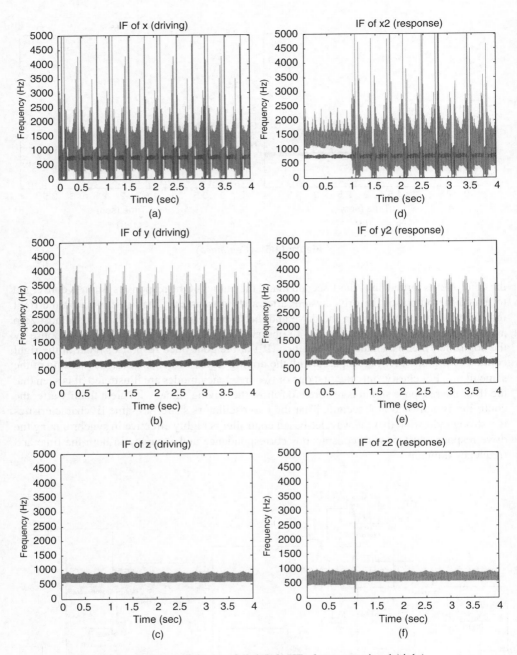

Figure 11.7   IF of driving signal (left); IF of response signal (right)

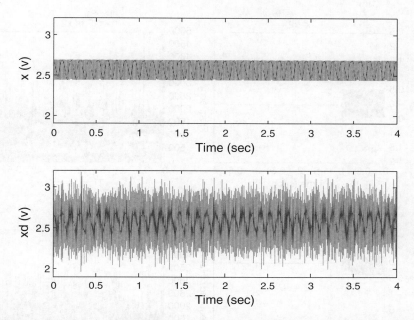

Figure 11.8   Signal x in the drive system and signal xd received by the response system

11.4.2   Scenario II

Synchronization of the first scenario is re-considered with a high frequency noise $d(t)$ being added to variable x during the transmission from the drive to the response system in Figure 11.3. The noise, which is a random signal of -0.5 to 0.5 volts in amplitude, is passed through a high-pass filter with a cut-off frequency at 4800 Hz. The x signal of the drive system is compared with the noise-scrambled signal received by the response system (xd) in Figure 11.8.

Figures 11.9(a–c) give the differences between variables in the drive system and those recovered in the response system. Even though the driving signal x is corrupted with the high frequency noise during transmission, the time-domain error between the driving and the response signals remains adequately constrained within a negligible range, a phenomenon understood as practical synchronization [3]. The frequency spectra in Figures 11.9(d–f) are restored to be of the same bandwidth as the driving system's in Figures 11.7(a–c), though not of the exact spectral characteristics of the driving signals. The result of synchronization not being affected by the transmitted noise indicates the level of robustness of the scheme of chaos synchronization.

11.5   Summary

Unlike other controllers, which focus mainly on the reduction of time domain error, the objective of the wavelet-based time-frequency control scheme is to mitigate the aberrant frequencies when the system undergoes nonstationary route-to-chaos processes. The parallel adaptive filter configuration allows on-line identification of unknown parameters without

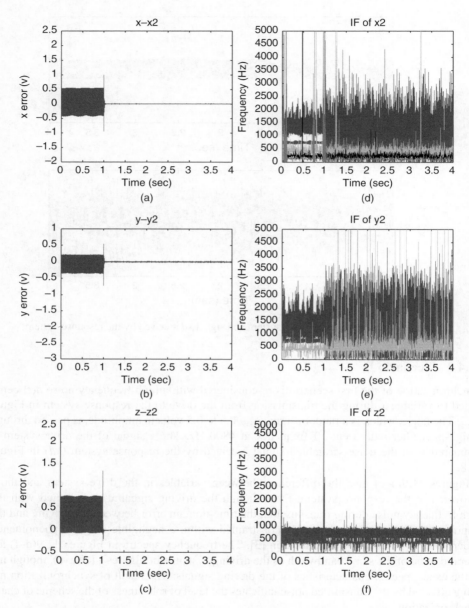

Figure 11.9 Difference between drive and response signals (left); IF of response signal (right)

resorting to closed-form linearization, hence preserving the inherent characteristics of a chaotic system and minimizing the impact of external disturbance as well as internal perturbation. As opposed to contemporary control practices, in which feedback loop is predominantly favored, the feedforward configuration prohibits the error from re-entering the control loop, thus reducing the risk of unintentionally exciting the sensitive chaotic system with adverse outcome. The fundamental conception of the controller dictates that, through manipulating wavelet coefficients, control is simultaneously exerted and achieved in the joint time-frequency domain. It is able to mitigate and properly restrain time and frequency responses of the chaotic system at the same time, regardless of the increasing spectral bandwidth that necessarily serves to invalidate and render ineffective common time- or frequency-domain-based controller design. This is also attested by the robustness demonstrated in effectively moderating the impact of high frequency noise.

Time-domain wavelet transform greatly reduces the computational load of the controller. As convergence time and time delay are expedited, the nonlinear effect of the controller itself is also minimized. The numerical study indicates that the nonlinear time-frequency controller not only synchronizes the nonautonomous chaotic systems in the time domain. It also recovers the underlying features of the driving signal in the frequency-domain without complete knowledge of the system parameters being available. The on-line identification feature of the controller allows the response system to start at arbitrary initial conditions and to be driven by sinusoidal forcing terms of different amplitudes and phases.

References

[1] Wiercigroch, M., Krivtsov, A.M., 2001, "Frictional Chatter in Orthogonal Metal Cutting," *Philosophical Transaction of the Royal Society*, A 359(1781), 713–38.

[2] Prian, M., Lopez, M.J., Verdulla, F.M., 2012, "Chatter Chaos Rejection by Adaptive Control," *AIP Conference Proceedings*, 1431, 676–83.

[3] Pecora, L. M., Carroll, T. L., 1990, "Synchronization in Chaotic Systems," *Physical Review Letters*, 64(8), 821–24.

[4] Femat, R., Solís-Perales, 1999, "On the Chaos Synchronization Phenomena," *Physics Letter A*, 262(1), 50–60.

[5] Chen, H. K., 2005, "Global Chaos Synchronization of New Chaotic Systems via Nonlinear Control," *Chaos, Solitons and Fractals*, 23(4), 1245–51.

[6] Park, J. H., 2006, "Synchronization of Genesio Chaotic System via Backstepping Approach," *Chaos, Solitons and Fractals*, 27(5), 1369–75.

[7] Yu, Y., Zhang, S., 2004, "Adaptive Backstepping Synchronization of Uncertain Chaotic System," *Chaos, Solitons and Fractals*, 21(3), 643–49.

[8] Feki, M, 2003, "An Adaptive Chaos Synchronization Scheme Applied to Secure Communication," *Chaos, Solitons and Fractals*, 18(1), 141–48.

[9] Chen, S., Lu, J., 2002, "Synchronization of An Uncertain Unified Chaotic System via Adaptive Control," *Chaos, Solitons and Fractals*, 14(4), 643–47.

[10] Lu, J., Wu, X., Han, X. and Lü. J, 2004, "Adaptive Feedback Synchronization of a Unified Chaotic System," *Physics Letter A*, 329(4), 327–33.

[11] Yau, H. T., 2004, "Design of Adaptive Sliding Mode Controller for Chaos Synchronization with Uncertainties," *Chaos, Solitons and Fractals*, 22(2), 341–47.

[12] Huang, C. F., Cheng, K. H., Yan, J. J., 2008, "Robust Chaos Synchronization of Four-Dimensional Energy Resource System Subject to Unmatched Uncertainties," *Communications in Nonlinear Science and Numerical Simulation*, 14(6), 2784–92.

[13] Park, J. H., 2006, "Chaos Synchronization between Two Different Chaotic Dynamical Systems," *Chaos, Solitons and Fractals*, 27(2), 549–54.

[14] Yassen, M. T., 2005, "Chaos Synchronization between Two Different Chaotic Systems Using Active Control," *Chaos, Solitons and Fractals*, 23(1), 131–40.

[15] Yassen, M.T., 2005, "Controlling Chaos and Synchronization for New Chaotic System using Linear Feedback," *Chaos, Solitons and Fractals*, 26(3), 913–20.

[16] Lü, J., Zhou, T., Zhang, S., 2002," Chaos Synchronization between Linearly Coupled Chaotic Systems," *Chaos, Solitons and Fractals*, 14(4), 529–41.

[17] Voss, H.U., 2000, "Anticipating Chaotic Synchronization," *Physical Review E*, 61(5), 5115–19.

[18] Pyragas, K., 1998, "Synchronization of Coupled Time-delay Systems: Analytical Estimations," *Physical Review E*, 53(3), 3067–71.

[19] Liao, X., Chen, Guanrong, 2003, "Chaos Synchronization of General Lur's Systems via Time-Delay Feedback Control," *International Journal of Bifurcation and Chaos*, 13(1), 207–31.

[20] Cao, J., Li, H. X., Ho, W. C., 2005, "Synchronization Criteria of Lur's Systems with Time-Delay Feedback Control," *Chaos, Solitons and Fractals*, 23(4), 1285–98.

[21] Yalçin, M. E., Suykens, J. A. K., Vandewalle, J., 2001, "Master-Slave Synchronization of Lur'e Systems with Time-Delay," *International Journal of Bifurcation and Chaos*, 11(6), 1707–22.

[22] Carroll, T. L., Heagy, J. F., Pecora, L. M., 1996, "Transforming Signals with Chaotic Synchronization," *Physical Review E*, 54(5), 4676–80.

[23] Carroll, T. L., Pecora, L. M., 1993, "Synchronizing Nonautonomous Chaotic Circuits," *IEEE Transactions on Circuits and Systems II: Analog and Digital Signal Processing*, 40(10), 646–50.

[24] Pecora, L. M., Carroll, T. L., Johnson, G. A., Mar, D. J., Heagy, J. F., 1997, "Fundamentals of Synchronization in Chaotic Systems, Concepts and Applications," *Chaos*, 7(4), 520–43.

Appendix

MATLAB® Programming Examples of Nonlinear Time-Frequency Control

In this Appendix two MATLAB® examples on the implementation of the nonlinear time-frequency control are listed. MATLAB® is a numerical computing platform developed by The MathWorks, Inc. that integrates visualization, programming, and the creation of a user interface in a highly open environment. The two examples all have a main program coded as an MATLAB® m-file that calls and invokes MATLAB® Simulink®. Simulink® is a graphical multi-domain simulation tool for the design and modeling of dynamic systems. Program initialization, parameter and coefficient assignment, and data transmission and storage are executed in the m-file. The system model with the "plant" that is to be controlled is constructed in Simulink® following the model-based design. The nonlinear time-frequency controller is developed in Simulink® and placed in the user-definable block called S-Function API. MATLAB® provides seamless signal transmission and storage between each subsystem and the controller. The implementation of the control of the friction-induced instability in Chapter 10 and the chaos synchronization in Chapter 11 are each demonstrated as an example in the following sections.

A.1 Friction-Induced Instability Control

Equations (10.5)–(10.18) in Chapter 10 are referred to. The finite element method (FEM) is employed to calculate the displacement of each elemental node in the cantilever beam which is in physical contact with the rotating disk. After the displacement field is determined, a finite-difference scheme is followed to estimate the corresponding velocity and acceleration of each node, as defined in Equations (10.15)–(10.18). The initialization of the coefficients, the configuration of the mass and stiffness matrices in the finite element model, and the construction of the wavelet transformation matrix are implemented in the main m-file. The corresponding Simulink® model is initiated by the main program and then the FEM calculation

Control of Cutting Vibration and Machining Instability: A Time-Frequency Approach for Precision, Micro and Nano Machining, First Edition. C. Steve Suh and Meng-Kun Liu.
© 2013 John Wiley & Sons, Ltd. Published 2013 by John Wiley & Sons, Ltd.

is completed at each time step. The architecture of the Simulink® model was given in Figure 10.9. The same figure is enlarged and reproduced in Figure A.3 for clarity at the end of the section.

A.1.1   Main Program

This main program is coded as an *m*-file as follows. The Simulink® system model "BEAM_5" is called by the main program.

```
================================================================
clear all; close all; clc;
% Assign model coefficients
N0 = 2;              % externally applied normal load
us = 0.4;            % the maximum coefficient of static friction
um = 0.25;            % the minimum coefficient of kinetic friction
vm = 0.5;            % velocity corresponding to um
vs = 10^-4;          % velocity corresponding to us
vd = 0.3;            % velocity of the disk
rm = 0.8;            % coefficient in eqn.(10.3)
tk = 100000;         % coefficient in arctan
f = [N0 us um vm vd rm tk vs];

% Assign coefficients for the finite element analysis
nel=10;              % number of elements
nnel=2;              % number of nodes per element
ndof=2;              % number of dofs per node
nnode=(nnel-1)*nel+1; % total number of nodes in system
sdof=nnode*ndof;     % total system dofs
tleng=1;             % total beam length
leng=tleng/nel;      % same size of beam elements
kk=zeros(sdof,sdof); % initialization of system stiffness matrix
mm=zeros(sdof,sdof); % initialization of system mass matrix
force=zeros(sdof,1); % initialization of force vector
index=zeros(nel*ndof,1); % initialization of index vector

% Assign simulation coefficients in Simulink
dt=10^-4;            % time step size
ti=0;                % initial time
tf= 200;             % final time
nt=fix((tf-ti)/dt);  % number of time steps
TC = 50;             % controller start time
fd = 0;              % desire tip position

% Construct system matrix (M, K, F) by finite element method
for iel=1:nel        % loop for the total number of elements
index=feeldof1(iel,nnel,ndof); % extract system dofs associated with element
```

```matlab
% Assign element mass matrix
m=rm*leng/420*[156    22*leng  54      -13*leng;...
        22*leng  4*leng^2  13*leng -3*leng^2;...
        54       13*leng   156     -22*leng;...
        -13*leng -3*leng^2 -22*leng  4*leng^2];

% Assign element stiffness matrix
k=1/(leng^3)*[12     6*leng  -12      6*leng;...
        6*leng  4*leng^2 -6*leng  2*leng^2;...
        -12     -6*leng  12      -6*leng;...
        6*leng  2*leng^2 -6*leng  4*leng^2];

kk=feasmbl1(kk,k,index); % assemble each element matrix into system matrix
mm=feasmbl1(mm,m,index); % assemble each element matrix into system matrix
end

% Add unit mass at tip
mm(end, end) = mm(end, end) + 1;
mm(end-1,end-1) = mm(end-1,end-1) + 1;
mminv=inv(mm);          % invert the mass matrix

% Initialize acceleration, velocity and displacement at current time step
acc = zeros(sdof,1);
vel = zeros(sdof,1);
disp = zeros(sdof,1);

% Initialize acceleration, velocity and displacement at next time step
acc_1 = zeros(sdof,1);
vel_1 = zeros(sdof,1);
disp_1 = zeros(sdof,1);

% Initialize tip frictional force by eqn.(10.4)
vr = vd - 0;
if abs(vr)< vs
force(end-1) = N0*(us*vr/vs) ;
else
force(end-1) = N0*tanh(tk*vr)*( us -3/2*(us-um)*( (abs(vr)-vs)/(vm-vs)-
1/3*(abs(vr)-vs)/(vm-vs))^3 ) ;
end

% force(1)= -1*force(end-1);
% force(2)= force(end-1)*tleng;

% Calculate the acceleration when the frictional force is applied
acc = mminv*(force-kk*disp);    % acc. = (f-kx)/m
acc(1) = 0;                  % transverse displ. at node 1 is constrained
```

```
acc(2) = 0;                  % slope at node 1 is constrained
ic = [disp vel acc];  % assign initial condition

% Construct the matrix of orthogonal WT by eqn. (7.12)
% Daubechies-2 (db4) wavelet
% Level 1 decomposition
N1 = 512; % length of the adaptive filter
h0 = 1/(4*sqrt(2))*[1+sqrt(3) 3+sqrt(3) 3-sqrt(3) 1-sqrt(3)]; % high pass filter
g0 = 1/(4*sqrt(2))*[1-sqrt(3) -3+sqrt(3) 3+sqrt(3) -1-sqrt(3)]; % low pass filter

% Construct wavelet transformation matrix
WT = zeros(N1);
for n = 1:N1/2
 if n == N1/2
    WT(n,N1-1:N1)=h0(1:2);
    WT(n,1:2)=h0(3:4);
    WT(n+N1/2,N1-1:N1)=g0(1:2);
    WT(n+N1/2,1:2)=g0(3:4);
  else
    WT(n,2*n-1:2*n+2)=h0;
    WT(n+N1/2,2*n-1:2*n+2)=g0;
  end
end

% Invoke Simulink
sim('BEAM_5');

% Display the simulation result
[m, n] = size(x);
time=0:dt:(m-1)*dt;
figure;
plot(time,Force)
xlabel('Time(sec)', 'FontSize',12)
ylabel('Force','FontSize',12)
title('Tip Frictional Force','FontSize',12)

figure;
subplot(2,1,1)
plot(time,x)
xlabel('Time(sec)','FontSize',12)
ylabel('Tip displacement (m)','FontSize',12)
% title('Tip','FontSize',12)

subplot(2,1,2)
plot(time,xd)
xlabel('Time(sec)','FontSize',12)
```

```
ylabel('Tip velocity(m/s)','FontSize',12)
% title('Tip','FontSize',12)

figure;
plot(time,vr)
xlabel('Time(sec)','FontSize',12)
ylabel('Relative velocity(m/s)','FontSize',12)
title('Relative velocity','FontSize',12)

figure;
plot(x(end/2:end),xd(end/2:end))
% title('Phase Plot','FontSize',12)
xlabel('Tip displacement(m)','FontSize',12)
ylabel('Tip velocity(m/s)','FontSize',12)
% ylim([-0.5 0.5]);
```
==

The following two MATLAB® functions are called by the main program above. They are adopted from the example code found in *The Finite Element Method Using MATLAB®*, by Kwon, Y. W. and Bang, H., 1997, CRC Press.

==
```
function [index]=feeldof1(iel,nnel,ndof)
%  Purpose:
%     Compute system dofs associated with each element in one-dimensional
%      problem
%
%     Synopsis:
%     [index]=feeldof1(iel,nnel,ndof)
%
%     Variable Description:
%     index - system dof vector associated with element "iel"
%     iel - element number whose system dofs are to be determined
%     nnel - number of nodes per element
%     ndof - number of dofs per node

edof = nnel*ndof;
start = (iel-1)*(nnel-1)*ndof;

 for i=1:edof
   index(i)=start+i;
 end

function [kk]=feasmbl1(kk,k,index)
%  Purpose:
%     Assembly of element matrices into the system matrix
```

```
%
%    Synopsis:
%    [kk]=feasmbl1(kk,k,index)
%
%    Variable Description:
%    kk - system matrix
%    k  - element matrix
%    index - d.o.f. vector associated with an element

edof = length(index);
for i=1:edof
 ii=index(i);
   for j=1:edof
    jj=index(j);
      kk(ii,jj)=kk(ii,jj)+k(i,j);
   end
end
```
==

A.1.2 Simulink® Model

Two customized MATLAB® function blocks, the finite difference function in Figure A.1 and the force function in Figure A.2, are introduced. The nonlinear time-frequency controller and the iteration of the finite element scheme are implemented in Simulink® as shown in Figure A.3

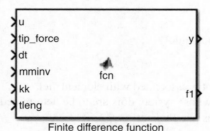

Finite difference function

Figure A.1   Finite difference function block in Simulink®

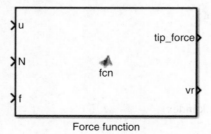

Force function

Figure A.2   Force function to calculate tip frictional force in Simulink®

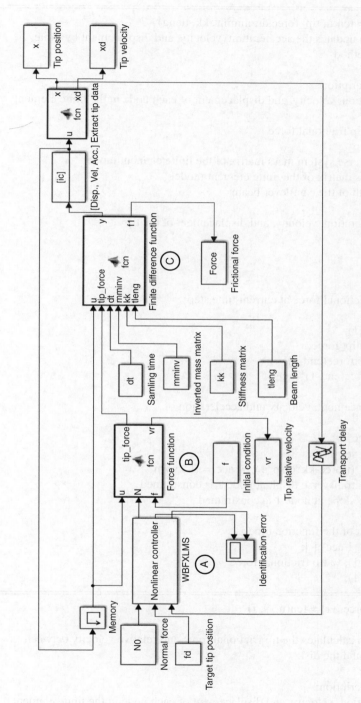

Figure A.3 Wavelet-based time-frequency control scheme of the disk-brake model

```
==================================================================
function [y,f1] = fcn(u, tip_force,dt, mminv, kk, tleng)
% This function updates the acceleration, velocity, and displacement by finite
% difference method
%
% Variable Description:
%   u - acceleration, velocity, and displacement of each node in the finite element
%       model
%   tip_force - tip frictional force
%   dt - time step
%   mminv - inverse system mass matrix of the finite element model
%   kk - stiffness matrix of the finite element model
%   tleng - length of the cantilever beam

% Extract acceleration, velocity, and displacement of the tip
acc = u(:,3);
vel = u(:,2);
disp = u(:,1);

% Assign tip frictional force at current time step
[m, n]= size(u);
tforce=zeros(m,1);
tforce(end-1) = tip_force;
%tforce(1)= -1*tforce(end-1);
%tforce(2)= tforce(end-1)*tleng;

% Update displacement, velocity and acceleration of
% the tip
vel_1 = vel + acc*dt;
disp_1 = disp + vel_1*dt;
acc_1 = mminv*(tforce-kk*disp_1); % acc. = (f-kx)/m
acc_1(1) = 0; % transverse displ. at node 1 is constrained
acc_1(2) = 0; % slope at node 1 is constrained

% Define output of the function block
y  = [disp_1 vel_1 acc_1 ];
f1 = tforce(end-1); % tip frictional force
x = disp_1(end-1);
==================================================================
function [tip_force, vr] = fcn(u, N, f)
%
% This function calculates the tip frictional force and relative velocity between
% the tip mass and the disk
%
% Variable Description:
%   u - acceleration, velocity, and displacement of each node in the finite element
```

```
%       model
%
%   f - [N0 us um vm vd rm tk vs]
%   N - normal force applied to the tip

% Extract tip velocity
% acc = u(:,3);
vel = u(:,2);
% disp = u(:,1);

N0 = f(1);    % externally applied normal load
us = f(2);    % the maximum coefficient of static friction
um = f(3);    % the minimum coefficient of kinetic friction
vm = f(4);    % velocity corresponding to um
vd = f(5);    % velocity of the disk
rm = f(6);    % coefficient in eqn.(10.3)
tk = f(7);    % coefficient in arctan
vs = f(8);    % velocity corresponding to us

% Calculate the relative velocity between the tip mass and the disk
vr = vd - vel(end-1);
%tip_force = N*(us*tanh(tk*vr)-3/2*(us-um)*(vr/vm-1/3*(vr/vm)^3));

% Update the tip frictional force according the relative velocity by eqn.(10.4)
if abs(vr)< vs
  tip_force = N*(us*vr/vs) ;
else
  tip_force = N*tanh(tk*vr)*( us -3/2*(us-um)*( (abs(vr)-vs)/(vm-vs)-1/3*(abs(vr)-vs)/(vm-
vs))^3 ) ;
end
```
==

A.2   Synchronization of Chaos

The implementation of the synchronization of chaos in Chapter 11 also consists of an *m*-file
and an associated Simulink® model. The initialization of parameters and coefficients and the
construction of the wavelet transformation matrix are implemented in the following main
program in the *m*-file syntax. The time-frequency controller and the chaotic system models
are built in Simulink® as follows.

A.2.1   Main Program

This main program is coded as an *m*-file, in which the Simulink® system model "Hyper_2" is
called.

```
========================================================================
clc; close all; clear all;
% Assign coefficients in the driving system in eqn. (11.1) to (11.3)
beta = 10000;
gamma_y = 0.2;
gamma_z = 0.1;
alpha = 0.2;      % amplitude of the sinusoidal forcing term
omega = 2*pi*769; % frequency of the sinusoidal forcing term
phase1 = pi/2;    % phase in the sinusoidal forcing term
int1=[0 0 0];     % initial position of the driving system


% Assign coefficients in the response system in eqn. (11.6) to (11.8)
beta1 = 10000;
gamma_y1 = 0.2;
gamma_z1 = 0.1;
alpha1 = 0.4;     % amplitude of the sinusoidal forcing term
omega1 = 2*pi*769; % frequency of the sinusoidal forcing term
phase2 = pi;      % phase in the sinusoidal forcing term
int2 = [2 2 2];   % initial position of the response system
%
Total = 4; % total simulation time
T = 1; % controller start time
Sampling = 0.0001; % sampling rate
%
N = 256;  % N: filter length
mu1 = 0.2; % step size of the LMS filter
mu2 = 0.2; % step size of the LMS filter
%
% Construct the matrix of orthogonal WT by eqn. (7.12)
% Daubechies-3 (db6) wavelet
% Level 1 decomposition
%
h = [0.3326705530 0.8068915093 0.4598775021 -0.1350110200 -0.0854412739
0.0352262919];
g = [0.0352262919 0.0854412739 -0.1350110200 -0.4598775021 0.8068915093
-0.3326705530];
% Construct wavelet transformation matrix
WT = zeros(N);
for n = 1:N/2
  if 2*n+4 - N == 4
    WT(n,N-1:N)=h(1:2);
    WT(n,1:4)=h(3:6);
    WT(n+N/2,N-1:N)=g(1:2);
    WT(n+N/2,1:4)=g(3:6);
  elseif 2*n+4 - N == 2
    WT(n,N-3:N)=h(1:4);
```

```
        WT(n,1:2)=h(5:6);
        WT(n+N/2,N-3:N)=g(1:4);
        WT(n+N/2,1:2)=g(5:6);
    else
        WT(n,2*n-1:2*n+4)=h;
        WT(n+N/2,2*n-1:2*n+4)=g;
    end
end
%
sim('Hyper_2'); % start Simulink
%
% Display the simulation result
figure;
plot(x3.time,x3.signals.values);xlabel('Time(sec)','Fontsize',12);
ylabel('x2(v)','Fontsize',12);title('x2','Fontsize',12);
figure;
plot(x3.time,x.signals.valuesx3.signals.values);xlabel('Time(sec)','Fontsize',12);
ylabel('x error(v)','Fontsize',12);title('x-x2','Fontsize',12);
figure;
plot(y3.time, y3.signals.values);xlabel('Time(sec)','Fontsize',12);
ylabel('y2(v)','Fontsize',12);title('y2','Fontsize',12);
figure;
plot(y3.time, y.signals.values-y3.signals.values);xlabel('Time(sec)','Fontsize',12);
ylabel('y error(v)','Fontsize',12);title('y-y2','Fontsize',12);
figure;
plot(z3.time, z3.signals.values);xlabel('Time(sec)','Fontsize',12);
ylabel('z2(v)','Fontsize',12);title('z2','Fontsize',12);
figure;
plot(z3.time, z.signals.values-z3.signals.values);xlabel('Time(sec)','Fontsize',12);
ylabel('z error(v)','Fontsize',12);title('z-z2','Fontsize',12);
%
a = 0; b = 0;
for i = 1:length(d.time)
a = [a d.time(i,1)];
b = [b d.signals.values(1,1,i)];
end
figure;
plot(a,b);
title('noise')
%
figure;
subplot(2,1,1)
plot(x.time, x.signals.values);xlabel('Time(sec)','Fontsize',12);
ylabel('x(v)','Fontsize',12)
axis([0 4 1.9 3.2])
```

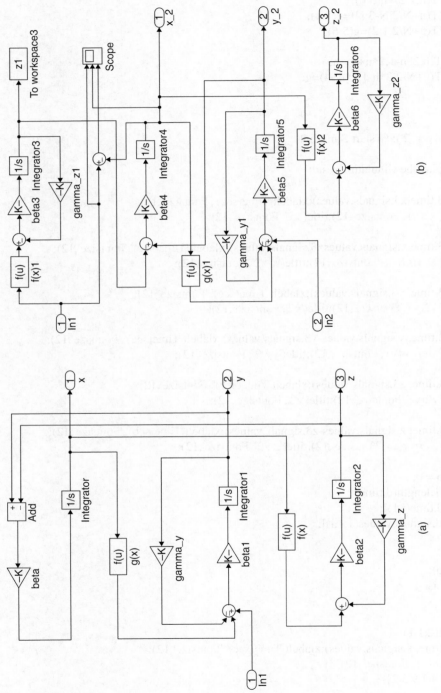

Figure A.4 (a) Driving system (b) Response system

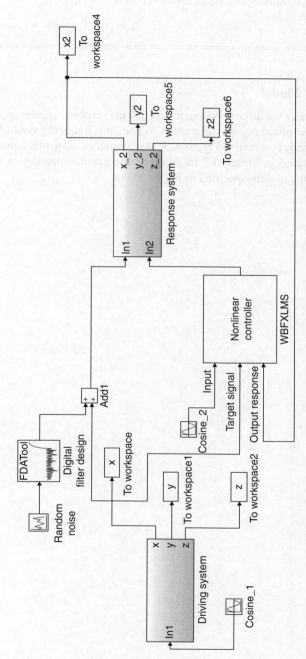

Figure A.5 Control scheme with wavelet-based time-frequency (WBFXLMS) controller

```
subplot(2,1,2)
plot(xd.time, xd.signals.values);xlabel('Time(sec)','Fontsize',12);
ylabel('xd(v)','Fontsize',12)
axis([0 4 1.9 3.2])
```
==

A.2.2 Simulink® Model

Two subsystems, namely the driving system model and the response system model, are built in Simulink®. They are explicitly given in Figure A.4. The time-frequency control scheme of the synchronization of chaos is shown in Figure A.5. The figures along with the provided function blocks can be duplicated in Simulink® to facilitate the synchronization of the drive-slave nonautonomous chaotic system specified in Figure 11.3.

Index

*Control of Cutting Vibration and Machining Instability: A Time-Frequency Approach for Precision,
Micro and Nano Machining*, First Edition. C. Steve Suh and Meng-Kun Liu.
© 2013 John Wiley & Sons, Ltd. Published 2013 by John Wiley & Sons, Ltd.